GHOST DETECTIVE

WRITTEN BY
MICHAEL J. WORDEN

WITH FOREWORD BY LINDA ZIMMERMANN

A SISU BOOKS PUBLICATION

To contact the author write:

Michael J. Worden
PO Box 302
Port Jervis, New York 12771

www.sisubooks.com
info@sisubooks.com

www.paranormalpolice.com
detective@paranormalpolice.com

Ghost Detective

Cover design courtesy of Linda Zimmermann.

Paranormal Police badge design by www.bluzdoggraphics.com

ISBN 978-0-9842283-0-0

DEDICATION

This book is dedicated first and foremost to my twin inspirations: Ryan and Michael. The challenges, rewards and joys of being your father evoke an emotion that I can not put into words. You inspire me to be a great father and a great person. I am honored to be your Dad. Thank you Kelly – for they are as much you as they are me.

To Linda and Bob: Thank you for your inspiration and friendship. Linda, your guidance and passion for the paranormal is amazing. Without your guidance this book would not happen. Linda, thank you for inspiring me and editing my work. Your encouragement made this book a reality!

To my family: for being supportive and as curious about the paranormal as I am. Especially Grandma – thank you for all that you do for me and the twins. We love you.

To Lisa Ann: The most gifted and talented psychic I have ever had the pleasure of working with. Thank you for keeping me motivated and staying on top of me to get this book completed.

To Barbara: a gifted psychic and good ghost hunting friend who is always finding wonderful places to investigate.

To the memory of Hans Holzer (January 26, 1920 – April 26, 2009): his books inspired me as a child and fueld my imagination. I began my journey as a ghost hunter in the pages of his books..

To my relatives who have passed on, your memory survives in our hearts. In particular to Grandpa, who was taken from us too soon. A day never passes where I do not miss you. My heart still aches from your loss.

To Judge O.P. Howell: the lost soul that dwelled in my grandparent's home for decades in search of his long deceased son, Bradford. May you have found the peace you deserve with your son.

Finally, I dedicate this book with a quote by the Roman philosopher Seneca: "Every new beginning comes from some other beginning's end". This is for all of the lost souls who are earthbound by choice or circumstance. Life is a precious gift to be lived, loved and cherished. My wish for you is to find that new beginning that comes from the end of your earthly life. Something better is waiting for you. For the living, live and love now and savor the gift that every new day really is. I often repeat a line from a song by the Icelandic band Sigur Rós "the best thing God has created is a new day". Truly, each new day is to be lived and savored. And for those who are curious about ghosts or enjoy looking for them as I do, never lose sight of the fact that these were once living people, lost now in the interzone between life and death. Respect them as you would the living.

CONTENTS

FOREWORD

In 1980 there was a rather ridiculous John Travolta movie called "Urban Cowboy" which spawned a craze across America. People who had never been within a mile of a horse suddenly bought cowboy boots and hats, and started line dancing and riding mechanical bulls.

What does all that have to do with a book about ghost investigations? In a similar fashion, the glut of paranormal television shows in recent years has led to an explosion of ghost hunting groups. Unfortunately, a clever acronym for a name, the bestowing of glorified titles, and silk-screened t-shirts do not a good ghost investigator make. In other words, these days everybody and their cousin seems to have taken up the mantel of paranormal sleuth, and hordes of these would-be ghost hunters have descended on every building, cemetery, or patch of ground which has ever had anything go bump in the night or sent the slightest shiver up an alleged eyewitness' spine.

In all fairness, this has been a good thing in many ways, as it has raised both awareness and acceptance of the possibility of the supernatural, and has produced some very competent and innovative investigators. However, it has also muddied the paranormal waters with those who exhibit considerably less-than professional conduct at a site, and then proceed to post all sorts of outlandish "evidence" on the Internet. For example, the curator of one historic fort had such a bad experience with an irresponsible group of ghost hunters that he has forever closed the door (or the gates, in this case) to any future investigations.

As trust and integrity are the keys to gaining access to homes, businesses, museums, etc., in the beginning I was something of a lone wolf in the ghost hunting world—I would rather go to a haunted site by myself than risk my credibility with people who were just their for fun and something to post on their websites.

Then about ten years ago I got an email from some cop named Mike Worden who knew of an old historic site that was supposed to have a few ghosts. I thought, "Hey, if you can't trust a cop, who can you trust?" so we arranged to meet at the site so we could check it out. Right off the bat I could see that Mike had great detective skills, was a first-class gadget man, and also had a unique sensitivity to paranormal energies. So I thought, "Hey, maybe it wouldn't be such a bad idea to have a cop at my next investigation."

Well, one investigation led to the next, then a few dozen more, and before I knew it there was Paranormal Police Detective Michael Worden with his impressive array of cameras and meters, all housed in custom-fitted cases, ready at a moment's notice to bring his expertise into haunted asylums, murder and suicide sites, battlefields, and more bizarre places then I can remember.

In all seriousness, I can honestly say without hesitation that having Mike as my ghost hunting partner was one of the smartest decisions I've ever made, and has led to more amazing adventures than anyone has a right to dare to ask for. Over the years we have laughed a lot, cried more than once, and yes, had the you-know-what scared out of us, but we always have each other's back and strive for the same goal—gather the best evidence we can to support the existence of ghosts. We take our work very seriously, but have never taken ourselves too seriously, and I think that combination of dedication and irreverent humor is what has made us click as a ghost investigating team.

It has genuinely been a pleasure and an honor to watch Mike's skills and common sense be applied to a wide variety of cases, and I am absolutely delighted that he decided to share his experience and wisdom in this book. If you want to be a ghost hunter—a good ghost hunter—read this book before you even think about going to a haunted location. Even if you already have a few paranormal investigations under your belt, this book can still teach you a lot about the theory and practice of ghost detective work.

Mike has always been humble about his capabilities, but he clearly has both the intellectual and intuitive gifts that make the ideal ghost investigator. But you don't have to take my word for it. His words speak for themselves.

Linda Zimmermann
The Ghost Investigator

INTRODUCTION

“Easy reading is damned hard writing”

- **Nathaniel Hawthorne**

Ghosts have fascinated me my entire life, a passion fueled by experiences at my grandparent's home. I was fascinated, scared and interested all at once. I read every book I could find on ghosts and watched the scarce television shows about them. I used to wonder why I was so fascinated by the paranormal. There was the typical adolescent dabbling in the paranormal: we were convinced that if you looked into a darkened mirror and said "Bloody Mary" ten times you would see a bloody, haggard face looking back at you. I am pretty sure it worked, and recall how more than one of us ran from the room describing the terror we had just witnessed! With the hindsight of time and age, I know that the power of suggestion, and of course, the fear of not wanting to be the only person never to see it, were at work here.

One of the stories I recall to this day is about a woman dying in childbirth. She was buried with her deceased infant in her arms. For several days after a pale woman was seen in town buying milk and walking off to the cemetery, only to vanish. Finally, out of shear curiosity, someone followed her one night and watched her vanish over the grave of the recently deceased mother and child. A speedy exhumation revealed that, while mom had passed on, the child had not and was still alive. Empty milk containers in the coffin verified that mom had returned from the grave to keep her child alive until help arrived. I never quite understood where this was supposed to have taken place. It just seemed to be somewhere local.

Then, of course, we had all heard about the phantom female hitchhiker who was taken to a dance, and her suitor, upon trying to locate her later, learns that she had died years earlier. We tried séances and Ouija boards. Once when we were young teens a group of us snuck into the basement while an older cousin and her friends were conducting a séance. We proceeded to make quite a ruckus and once the screaming in terror died down they chased us down the street yelling and cursing at us!

In 1987 I spent two weeks with friends in San Antonio, Texas. In addition to a moving and reverent visit to the Alamo, there were the ghost stories. All of them true. There was a set of railroad tracks where it was said that a school bus full of children died in a tragic train accident. I was told, and believed, that if you put baby powder on the back of your car and stop your car at that very grade crossing, phantom hands will push you off. As if that were not eerie enough: phantom handprints would be left in the powder.

Urban legends, or ghost stories, are a part of our culture. I have heard my own nieces talking of phantoms in mirrors. Most places have a phantom railroad tunnel, hitchhiker, or spirit that wanders a graveyard. Stories such as these helped to drive my desire to know more. But growing up in the 1980s, it was not your typical topic of conversation. So I buried myself in books by Hans Holzer, Dan Cohen and whomever else that I could find in the library. I watched *Ghostbusters* on VHS so many times that the tape began to disintegrate. I knew that being a spectator to the field of the paranormal was not going to be enough for me.

My first excursion into the world of ghost investigation was around 1990. My brother Scott and I set up a small, micro cassette recorder in one of the bedrooms at my grandparent's home while they were away. We selected the room because it seemed to the most active on the second floor. We set the recorder and left for the night, allowing the recorder to do its thing.

The next morning we retrieved the tape and listened to the recording. We were astonished. It was clear that you could hear doors opening and closing, someone walking in the hallway outside of the bedroom, footfalls up and down the stairs. Even in the distance you could hear sounds like the kitchen cabinets slamming shut. The squeaking sound of the handrail on the steps was clearly audible more than once. Something had been in the house making noise and we had captured about an hours-worth on tape.

We ended up doing this more than once, each time capturing unexplained sounds in an empty house. Sadly, those tapes are lost to time. My first attempt at ghost hunting may have been slightly clumsy, unorganized and awkward, but the basis was there. I wanted to know more. And I was going to find out.

In 2001 I was fortunate to begin working with author and paranormal researcher Linda Zimmermann. I am not sure how that all came about, but suffice to say, we ended up making one hell of a paranormal team. This book is as much about her work as it is mine. Without her guidance, patience, and molding I would not be the paranormal investigator I am today. Throughout this book you will see references to her and our investigations and paranormal experiences. Her *Ghost Investigator* series of books chronicles and details our paranormal escapades and if you have not yet read one, do so. But be warned: you will find that you can't just read one of her books. You will want more.

This book has been a work in progress for some time now. That progress has often been slow - very slow. I did not want to write a book of just ghost stories or chronicles of investigations. Linda does that extremely well. I did want to write a dry, technical how-to-manual either. I have seen popular television shows where new investigators have to receive extensive training on using EMF meters or non-contact thermometers. Believe me; you do not need a manual to learn how to use those things.

Time, and the slow rate of progress, have shaped and defined the book that you are reading. It is the experiences of a ghost investigator – with some how-to and good old fashioned advice mixed in. I offer a unique approach in that I bring my experiences as a police officer to ghost hunting. This book is about a field of research that is growing and defining itself. In 2001 ghost hunting was an almost unheard of field of science. Ghost hunters were odd people that chased ghosts using tools that made them comparable to a character in *Ghost Busters*. People who investigated ghosts faced more ridicule than respect or understanding. Things have made a dramatic turn. Popular television shows have turned ghost hunting into a growing field of study. Ghost hunting groups have sprung up all over the United States and indeed, other parts of the world. Several cable networks now have reality-type programs featuring ghost hunting. Talking about ghosts in a serious manner is no longer frowned upon. It is becoming the norm.

As with any emerging field there are different theories of practice and methods of seeking and evaluating evidence. Linda and I have always worked hard to refine our own, unique approach to ghost hunting and investigating. We have worked hard to maintain a high level of credibility, integrity, and professionalism. We do this while keeping grounded and maintaining not only a sense of enthusiasm, but a sense of humor as well.

Whether you are a seasoned investigator or a novice to the field, I hope that are able use the information to make you a better ghost hunter or at least, a better consumer of ghost hunting information.

The author pictured taking a break during one of his first investigations. Photo taken at Smalley Inn, Carmel, New York. Courtesy Linda Zimmermann

There are a lot of opinions, none of them necessarily right or wrong. Read and digest all that you can. In the end you, too, will be a ghost hunter.

A word on terminology: I have used many terms in this book interchangeably: ghost hunting, ghost investigation, paranormal, haunting, activity. It can be argued that ghost hunting is different than investigation. Ghost hunting is often used to describe going to a place without prior paranormal reports and looking for ghosts, such as a cemetery. Ghost investigation implies investigating reports of the paranormal at specific locations. Most of my experience has been in the investigation of alleged hauntings. However, in this book the term ghost hunting and ghost investigating will be used interchangeably.

Finally, to the skeptics I offer this selection from Hans Holzer's *Real Hauntings: True American Ghost Stories:*

> "...I would suggest to my readers not to argue the existence or non existence of ghosts and haunted houses. Everyone must find his or her own explanation for what he or she experiences, and belief has nothing to do with it. Belief is the uncritical acceptance of something you cannot prove one way or another and the evidence for ghosts and hauntings is so overwhelming, so large, and so well documented, that arguing the existence of the evidence would be a foolish thing indeed. While there may be various explanations for what people experience in haunted houses, no explanation will ever be sufficient to negate the experiences themselves."
>
> - Hans Holzer
> January 26, 1920 – April 26, 2009

Now delve into the world – my world – of ghosts and the supernatural. I hope you enjoy reading this book as much as I have enjoyed writing it.

Michael J. Worden
Summer 2009

THE NATURE OF GHOSTS AND HAUNTINGS

"One owes respect to the living. To the dead, one owes only the truth."

- **Voltaire**

What is a ghost? Seems a rather simple question, but in reality it is much more complex. It is a question that goes to the heart of what we do. Ghost hunting is, after all, an effort to prove the existence of ghosts. And if you are going to go out and look for them, you should at least have a pretty good idea of what you are looking for. First I want to tell you that ghosts and hauntings are two different things. Often they can occur together and become interchangeable terms. But in essence a ghost and a haunting can be very different. A haunting is the experience of paranormal activity. It may be caused by an actual ghost or spirit, or a psychic imprint. This is where the difference is found.

It is difficult to venture into a discussion on the nature of ghosts without treading upon religious views and beliefs. Different religions view the soul and life after death in various contexts. Suffice to say, I am going to avoid religious overtones and look for a practical explanation of ghosts. A ghost is the remnant of a once living human. It is the spirit, essence or personality of that person that has survived the death of the physical body and now remains here at this level of existence. There is something within us that transcends the physical body and survives after our death. When that stays here on earth, it is a ghost.

If a ghost is a part of us that survives death but doesn't move on, what then is a haunting? We generally find that hauntings fall into two broad categories: residual and intelligent. A residual haunting, or psychic imprint, is where something (such as a person or event) has imprinted itself onto the environment much like a video signal is imbedded on a video tape. Residual hauntings are like a recording. They play back under certain circumstances, but you can not interact with them and they can not interact with you. A ghost, on the other hand, is dynamic and interactive. You can interact with a ghost, and a ghost can interact with you. These we can consider intelligent hauntings.

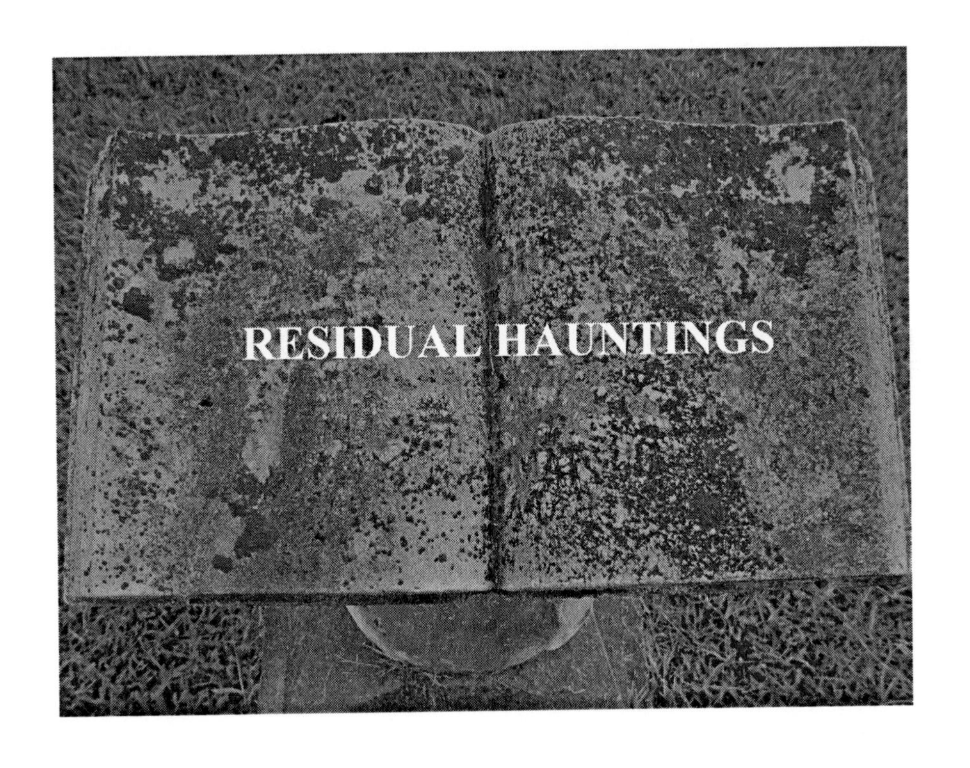
RESIDUAL HAUNTINGS

A residual haunting is comparable to a video tape. If you record an event on a camcorder and then remove the tape and set it on the table, all you have is a videocassette. The image is imbedded on the magnetic medium of the tape. But you can not see it. People can walk by the tape, even pick it up and look at it, but they will not see what it is that you have recorded on it. However, if you have the proper playback mechanism (e.g. a VCR) then you can insert the tape and play it on your television. Suddenly the event comes to life on your screen. You can watch it but you can not change or influence what has already been recorded. You can yell at the images but they will not answer you back. It is a recording. It is unalterable.

Now imagine traumatic events occur at a location: a murder, suicide or tragic accident, for example. The energy of the event may imprint itself on the location. Now the imprint is stored there much like the images on your video tape. Under the right circumstances those imprinted images are visible to people and we get to experience the events – or portions of them – that have been recorded on the environment.

What are these circumstances that allow us to access residual hauntings? For some it is the person itself. Some people are more sensitive to them than others. That may explain why some people can access residual events while others can not. I like to compare it to trying to put an 8mm video tape inside of a VHS player. It just doesn't work. Put that same 8mm tape into an 8mm player and now you can watch the video.

I also believe that for some residual hauntings timing is another critical element. Certain hours of the day, specific days of the week or the year, or events may be necessary to access these imprints. For example the anniversary of a death or a murder may be the trigger for accessing that residual haunting. Perhaps a battle replays itself once a year on the anniversary of the battle. Or a lonely widow pining away her last days pacing the floor at night may leave an imprint that occurs only at night at the times she would have been pacing.

A great example we have experienced is the Van Winkle House in Hawthorne, New Jersey. Here in the early hours of January 9, 1850, a laborer named John Johnson climbed into a second floor window of the residence of Judge John Van Winkle. After enjoying some fresh mince meat pies, Johnson entered the Van Winkle's bedroom and attacked the Judge and his wife, Jane, with a hatchet. Mrs. Van Winkle died quickly. Judge Van Winkle was able to survive long enough after the assault to name his assailant as well as plan his funeral. Johnson would later hang for his crimes. The crime scene itself was closed off and remained sealed up until it was opened in 1882, still containing blood and evidence from the crime.

Linda and I had the opportunity to brave an icy January 2005 evening to investigate the Van Winkle house on *the anniversary of the murders!* While set up in the room where the murders had occurred at around the same time it was believed that the crime was committed, Linda and I both heard a high pitched wheezing sound in the room that sounded like someone taking a dying breath. We captured it on two different camcorders that were running. It was a startling moment that leaves you stunned. As investigators we had gone to this place on the anniversary night hoping to experience something. Was this experience the final agonizing moments of Judge and Mrs. Van Winkle imprinted on the room? Had we accessed that terrible moment and experienced a part of it?

A residual haunting is experienced. You may see, hear, smell even feel the events. You may smell perfume or tobacco smoke. You may see what looks like the apparition of a person. You may hear foot steps or talking or music. You may experience the event as if it were very real and very palpable. But the event simply replays itself with out any changes. It is the same thing every time. You can not talk to the apparitions, or influence what they are doing. They will do the same things over and over again, time after time, in the same manner as events recorded on video tape.

The apparitions that you see in these residual hauntings, that is the images of what appear to be ghosts, are not ghosts. They are not the spirits of a dead human being. The ghost is not actually there present in the room. It is only the imprint of a once living person. It can not hear you, see you or interact with you. And you can not interact with it. These residual hauntings are also very specific to locations. An event imprinted on a second floor hallway will not wander down to the ground floor kitchen, for example. While there may be other layers of imprint in that hallway, or even in other places at a location, an imprint is not mobile. It does not move from room to room. It stays where it was imprinted.

How do you know when you are encountering a residual haunting? There are no fast and hard rules for this. Ghost hunting is as much art as it is science. However, as part of your investigation, get good background information on the location. Find out what people have encountered. If it seems to be the same thing over and over again, especially at similar times, then you may have a residual haunting. If people tell you that they see an apparition but it never reacts to them, then think residual. Do certain things only seem to happen at certain times or specific places? Are there documented events, especially tragic or traumatic ones that can be tied to the location? Ultimately you make a decision based upon the totality of your investigation as to whether you believe it is residual in nature.

INTELLIGENT HAUNTINGS

Far more common from my experiences are the intelligent hauntings. These are the rather close encounters with the spirits of deceased people. These encounters can range from the rather comfortable to the very frightening, depending upon the personality of the person who has remained behind. And what are the reasons that a ghost would remain here? I believe there are several reasons that ghosts remain earth bound and they generally fall into two broad categories: involuntary and voluntary.

Involuntary ghosts are not here by their own choice. In fact they may not even be aware of where they are or what has happened to them. What types of events can lead to an involuntary ghost? Sudden and unexpected deaths, especially those due to trauma, seem to be major causes. These are instances where the deceased had almost no time to adjust to the sudden transition from living to dead. There was no time for the spirit to comprehend what was going on and what was happening to him or her. The sprit thus becomes stuck here, at this level of existence. Referring back to the experience of the death breath at the Van Winkle House – it could be possible that Mrs. Van Winkle, who was murdered brutally in her sleep, had no opportunity to transition. One moment she was sleeping, the next she was brutally assaulted. Even if she had awoken for a moment she most likely had little or no understanding of what was happening to her. Mrs. Van Winkle could be quite the candidate for a ghost.

These unfortunate souls may not even know that they have died, let alone understand where they are or what is happening to them. They may become stuck at the place of death or even at a place they were attached to in life.

Traumatic events, such as a violent homicide or car accident for examples, may result in a stuck spirit. Anything where a person dies suddenly and without warning can be seen as good circumstances under which a ghost may end up wandering about. Violence seems to accompany these events as sudden death is often precipitated by violent acts or accidents. However, don't assume that a confused and lost soul is automatically the victim of violence. There are medical reasons where apparently healthy people die suddenly and with out warning, such as in sudden cardiac death, aneurysms and strokes.

How common are involuntary ghosts? We have encountered them on investigations, but I do not think that they are very common. If you do have the opportunity to interact with one of these lost souls, take the time to explain to them about where they are and what has happened to them. Encourage them to assess their situation and to move on. I tell spirits to look around and draw upon people that they loved and were connected to in life. I tell them to look for loved ones who have already crossed and to ask for help. It is waiting for them and they do not have to be stuck and confused. Be firm and compassionate with them. Be empathetic. Remember that you are talking to the spirit of a person who once lived just as you do. Someone who loved, laughed, cried and lived. And that person is stuck somewhere. They are alone, afraid, confused, lost and uncertain. They may be clinging to things that are familiar for a sense of security and understanding. If this were a loved one of yours you would want them to move on.

Stay behinds are a more common form of intelligent haunting. The stay behind is a ghost who is here by choice. The ghost knows that they are dead and free of the physical body, but it does not want to move on. Something keeps the spirit here, earthbound.

What circumstances could cause a spirit to willingly stick around after death? The reasons are as diverse and unique as people are in life. Issues such as attachments, fears, unfinished business, grief, and emotions are examples of reasons a spirit may voluntarily stay behind.

Often ghosts seem to be attached to some element of life. Perhaps they are attached to their home, or their favorite room in home. They may be attached to objects or locations. One theory regarding cemetery hauntings is that some ghosts may simply remain near where their physical body is buried. Attachment to their earthly remains – the reminder of what they once were – may be something that keeps them there. Some spirits may even be attached to the living that they leave behind and simply chose to stay around as not to miss out on anything.

Fear seems to be another and we have encountered a few that could fall into this category. What could a spirit fear that would keep them bound here on earth? Fear of punishment. I mentioned at the beginning of this chapter that I was going to avoid religious overtones. Here is where religion may play a role in hauntings and one which we have had experience with. Suicide is a painful experience for any family to go through. It has been a societal and religious taboo that still forces many families into shame. For many with strong religious views suicide is a grave sin. We investigated a home near Port Jervis, New York, where an elderly individual suffering from cancer had killed himself at home during a break in hospitalizations. The man had strong religious views, yet his illness drove him to the point of despair where suicide was a better alternative than life suffering from a terminal illness. The man's presence was felt and seen in the home, and the spot where he had died was particularly active. Can the fear of punishment keep a spirit from moving on? If someone believes that suicide will lead to damnation of the soul, can that serve to keep them from moving on? How about someone who believes certain sins can damn the soul to hell or some other eternal punishment? Perhaps the fear of damnation can cause a spirit to remain after passing - fearful of a perceived punishment.

Perceived is a word that I have chosen intentionally. I am writing this with a neutral view on religion. It is not necessarily reality that is as important as is what the spirit believes at the time he or she passes. No matter what the reality is – and none of us truly know that until we die and pass on – a spirit firmly believing something will carry that belief and it may influence their ability or willingness to move on after death. No one wants to be punished and certainly fear of eternal punishment or torment may be sufficient to keep some ghosts around.

Unfinished business is another reason ghosts stick around. Perhaps it is to witness a birth, marriage or other event. It may be that they are looking for something. One spirit that we encountered was that of Judge O. P. Howell, a former Surrogate Judge of Orange County, New York. He is a ghost who I feel an affinity for as he is a spirit who haunted my grandparent's home for the better part of a century.

My grandparent's home dates to the 1850s, has been in my family since the 1950's, when my grandparents bought it and set about turning it into a home where they would raise my father and entertain four grandchildren, and now, five great-grandchildren.

What would unfold from our investigations and research at my grandparent's home was a story of a tragic loss and life-altering grief and sadness. A grief so powerful that it was able to reach out across a century, and transcend the barrier between the living and the dead!

Obadiah Pellet Howell, former Surrogate of Orange County, New York, and descendant of Governor Bradford of Plymouth Colony.

Growing up, we spent a lot of time at grandma and grandpa's house. We would often spend the night and camp out on the living room floor. These nights were often filled with fear as I would lie on the floor and hear the sound of footsteps on the staircase not more than fifteen feet away. I would also see what appeared to be someone in the area of the bay window in the dining room, and feeling uneasy, turn my back to the doorway that joined the living room and dining room.

I used to walk to their house after school and wait for them to come home from work. One afternoon when I was in my early teens, I was watching TV in the living room. Without warning, there was a loud crashing sound in one of the upstairs rooms. It sounded like someone knocking over a pile of boxes in the back room of the second floor. I ran out of the house and did not go back in until grandpa had come home and he checked to make sure that no one had burglarized the place. They hadn't.

I can also recall numerous instances of being in the front bedroom doing homework. There was a desk there and it used to face the wall. I would only have to sit for a few minutes before the overwhelming feeling that someone was standing behind me would compel me to spin around, only to find an empty room.

I was not alone in these feelings. One summer after I had graduated high school, I spent a week watching the house while my grandparents were away. One of my cousins had stayed the night and we had ended up crashing in the living room. This particular cousin was related by the other side of the family and had not spent a lot of time in the house. The morning after his first night there he complained about the walking in the second floor hallway and on the staircase. I recall how he described it as someone walking from the bathroom area to the steps, down the steps, back up the steps and back towards the bathroom.

Many others had unearthly encounters over the years. One of the stories told to me by my mother is that during a party a man was seen to enter the front door and walk upstairs to the second floor, only to vanish. Now to understand this situation let me describe the house. The kitchen area is partially open to the dining room which is open to the living room, where the front door and staircase are located. You can stand in parts of the kitchen and see across the house to the living room and front door.

As mom described it, a man in a trench coat and hat entered the front door and proceeded to walk up the steps to the second floor. It seemed odd, I am told, as most people used the back door to the kitchen to come and go. But there was a border living at the house, Ralph, and he may very well have come in. However, when Ralph walked through the back door not long after it became clear that some else had come in the front door. A check of the second floor did not find anyone.

My grandmother also told me a story about being on the second floor cleaning one afternoon when a man proceeded to walk up the steps and go into the bathroom. As she has told the story, she hid until she heard the man leave the bathroom and proceed down the stairs and out the front door.

Having experiences in my own childhood, it is not surprising that other children have had encounters with something paranormal in the home. My brother's daughter Brianna was four years old and playing in the front bedroom with some dolls when she came running down the stairs and said that there was a man in the mirror. She described him as an older man wearing a hat and coat. The room was checked but no one saw anything unusual.

Brianna witnessed this apparition from the mirror again on another occasion. This time I was present along with my sister Melissa. We were in the living room with Brianna when she pointed to the bottom of the steps to tell us that there was a man there. She was insistent and startled by this figure—a figure that Melissa and I could not see. I picked her up and she clutched tightly against me. Whatever Brianna was seeing was not only very real to her, it was very frightening. She explained that the man had gone up the stairs so we followed. Brianna said that the man had gone into grandma's room, which is the middle bedroom on the second floor. We entered the room and Brianna pointed to the corner to show us where the man was standing. While the apparition of the man scared her, probably most frightening part of all this to her four year old mind was the fact that Melissa and I could not see him. Brianna asked aloud more than once why we couldn't see the man.

Was this apparition the same that was seen decades earlier by my mother? It very well could be. And he was far from done making appearances in the home. My own twin sons had their encounters with the apparition of a man. When they were two, I was making them breakfast while Ryan watched TV in the living room and Michael played in a toy room off of the dining room. In the middle of scrambling eggs, Ryan came running into the kitchen very upset, telling me about a man and pointing to the front of the house. I picked him up and he hugged me tightly, still repeating that there was a man. I brought him into the living room, double checked the front door (it was locked and dead bolted) and set him down to finish breakfast.

A few minutes later, Ryan ran back to me in the kitchen and grabbed onto my legs, telling me about a man and pointing to the living room. I carried him back in to the living room, explaining to him that there was no man. This time Ryan pointed to the steps and slowly followed something up the stairs--a man he told me. About a half hour later I was in the living room with both of the twins when they both became agitated and upset, pointing this time to the back door and telling me about a man.

Another appearance of the man was a month or so later. Their cribs were in the front bedroom, and the twins had just had a nap. This is the same bedroom where my niece, Brianna, had seen the man in the mirror. In this encounter, both twins *insisted* that there was a man in the mirror. Another similar encounter occurred when I was putting them to bed. A few minutes after setting them down for the night, I heard them screaming at the top of their lungs for daddy. As a parent this is one of the most unsettling sounds – that of your children in trouble. And that is what it sounded like.

I ran into to the room to find them standing in their cribs, crying, shaking, and pointing to the corner of the room. They were yelling to make the man leave the room. I looked to the corner and saw no one and tried to calmly explain that there was no one there. The twins insisted that there was a man there and told me to make him leave. Seeing my children upset cast an uneasy feeling over me, but something else in the room seemed to be influencing me. Something felt "off" in the room and I asked the twins what the man was doing. Their answer, given at the same time, was that he was kneeling down and praying!

Not all personal experiences were as terrifying. Once I had come home after working the overnight shift and checked in on the twins. Ryan was awake so I changed him and lay down on the bed in their room to wait for his brother, Michael, to wake up. After a few minutes Ryan began pointing to the mirror. After doing this several times he pointed to the closet. Then he pointed to the mobile above his crib, *which was spinning slowly.* Now I had fastened the mobiles to the ceiling above their cribs, and at a height that I could stand under them with out hitting them. Also, Michael's mobile was completely motionless.

The twins also had repeated encounters with something more sinister, at least in description. I would often hear from them that there was a man in their closet. They would sometimes call it a creature. When pressed for details they always said it was a man, all black, and he had no arms or legs and just crawled around on the floor from the closet!

I myself, looking back, never liked staying in that front room and recall that in 1988 when my grandfather was in the hospital for a heart problem, I stayed there and slept with a light on for the entire week. I was 18 years old! In hind site this was not the best room for my sons to sleep in!

I have had more recent experiences, as well, including one in the spring of 2009. I was sleeping in the middle room when I stirred in the night, opened my eyes, and saw an old woman standing in the corner of the room. She was gray haired, with her hair pulled back, wearing a long dress, and she was gazing out into the hallway. No sooner did I see her than she faded into the darkness. Another experience a few years earlier occurred when I had woken one night to check on the twins and use the bathroom. From the bathroom I watched as an apparition of a dark figure stood in the bay window on the second floor then walked into the back room, where I was sleeping!

The investigations at my grandparent's home were never uneventful. During one investigation Linda and I were in the front bedroom and telling the spirit to move on. At this point we did not know about Judge Howell so we were not speaking directly to him. Linda and I both experienced a life altering sadness come over the room. It was a sadness that elicited tears and a visceral sadness. At the time of this experience I had not yet become a father. Despite that minor fact, I described the feeling as being a life altering sadness so great it could only be compared to the sadness one must feel at the loss of a child.

My grandmother and I were motivated to do some research on the previous occupants of the house and identified Judge Howell as being one of them. During a visit to the local historical society we obtained the obituary for Judge Howell as well as his burial location in a local cemetery. We also managed to locate a photograph of the Judge. Curiously, when the old volume was taken off of the shelf the historical society volunteer opened it up to the exact page we were looking for – the page with the photograph of O.P. Howell!

The picture of O.P. Howell resembled the apparition that had been seen by many in the residence over the years. But a resemblance to a photograph would be minor evidence compared to what we would learn from the obituary and burial site.

Years earlier while patrolling through the cemetery something had banged on the roof of the patrol car. There was nothing to account for the banging on the top of the car and I always kept a mental note of the location. When I obtained the burial location and went to check on the site, the Howell family was buried at the exact location where something banged on the roof of the patrol car.

The Howell Family plot in Laurel Grove Cemetery.

More fascinating is the information we learned from the obituary of Judge Howell. The following excerpt is from the April 28, 1909 obituary:

"His only Son Bradford Howell, named in honor of the family's distinguished ancestor, died at an early age and the father never fully recovered from the blow." *Never fully recovered from the blow*? Months earlier Linda and I had encountered a presence with the overwhelming life altering sadness – one that we both felt was the loss of a child! Had we tuned into the sadness of a father grieving for his son? Was the restless spirit in my grandparent's home stuck there looking for his long dead child?

Bradford Howell's short life began on July 11, 1885 and ended from diphtheria on November 12, 1892. The loss of his son must have been a *life altering* occurrence for the Judge, as it was important enough to write about in his obituary 17 years later. Even Bradford's grave is unique in the family plot, and an inscription on the main family stone mentions that he 'fell asleep'.

Back of the Howell Family main stone with the inscription of an unnamed child that passed March 14, 1871, and the inscription for Judge Howell's beloved Bradford with the date he 'fell asleep'.

The inscription on Bradford's headstone is fading. It reads: 'He carries his lambs in his bosom'.

Armed with fresh information we were able to conduct another investigation, this time a more personal level. We took the time to speak directly to Judge Howell and to tell him that Bradford was not in the house and that he could be with Bradford whenever he wanted to. We told him that Bradford was waiting for him – to reach out to him – and that it was ok to move on. Our tone was encouraging and empathetic. We wanted Judge Howell to find his lost son on the other side. I believe our persistence worked. The presence that was once in the house has changed. Yes there are other spirits still in the residence (interestingly enough, where we find one ghost, there tends to be others) but those do not seem to have the personal connection to the residence that Judge Howell had in his century long quest to find his long dead son. Looking at the totality of the history of haunting activity and our research and investigations, Judge Howell may have been the best example of a stay behind that we have encountered. He was able to move on once he realized that he was not going to find Bradford here at this home and at this level of existence. Free from his earthly quest I am sure that he found his son waiting patiently on the other side for one touching and long overdue family reunion.

What other reasons could lead a ghost to stay behind? An unlimited number of reasons – reasons as unique as we are. For we are all potential ghosts. By that I mean that each one of us has the potential to remain here after the death of our physical body. If you chose to live a life of regret, or fear, or sadness, or carry around heavy burdens, then you may be setting yourself up to become a ghost. Live life. Savor life. Love life. It is short, fragile and often ends abruptly and without warning. I have seen that too often in my role as police officer. None of us are guaranteed anything, let alone the knowledge of how long we will live and when our life will end. As a police detective I see the tragic side of life and the pain and suffering that people experience, and bring upon themselves and others.

As a ghost investigator you will encounter restless souls who are stuck here for an infinite number of reasons. Each reason as unique as the ghost itself. The diversity of ghostly personalities is no different than the diversity of those personalities in life. On investigations we get caught up in the experiences and the activity. We focus on the sounds or the EMF readings or the temperature fluctuations. All of that is well and good. But sometimes answering the 'who' and 'why' is equally as important.

More than once we have encountered a haunting with a story to be learned. These hauntings tend to be very personal and very intimate encounters. Common among these encounters is that once we have conducted our investigation and pieced together the story, we learn that things have quieted down. The spirit of Judge Howell is a great example of this. Perhaps some of those restless spirits are simply trying to get someone to listen to them. Once that has happened they either move on to the next level of existence or, at the very least, stay around but stop trying so hard to be heard.

What is the story that some ghosts want to tell? That is not a question with a simple answer. For each ghost encountered will have a story or a secret as unique as the ghost itself. It becomes for us – the investigators – to try and uncover their story. In this way we truly owe the dead, or at least the restless dead, the truth.

There are other types of hauntings that I will not be considering here in this chapter. Demonic possession and poltergeists are two of those instances. I do not doubt that these hauntings are legitimate. People can be, and are possessed by, any manner of ill spirits, not necessarily demonic ones. Quite often the living people at a haunting are generating or feeding the activity. To borrow a phrase from Linda Zimmermann 'hauntings are as much about the people as they are the ghosts'. My advice to those living in a haunted location is to take control of the haunting. Do not react to every sound or every shadow. Do not relinquish control over your own home, and do not make the haunting the center of your world. I often like to compare ghosts to a child throwing a temper tantrum. If you ignore the child he or she will stop kicking and screaming. I have twins, I know something about tantrums! But if you try and intervene and yell at them in the midst of a fit it only will add to it and cause the tantrum to worsen. The same can be said of some ghosts. The moment you begin to recognize their presence they may draw upon that and continue to be pester you. It is ok to recognize them – my grandmother still yells at them in the night to be quiet if they are getting too loud – and try and find out what they want or why they are there. Just don't make them the center of your world.

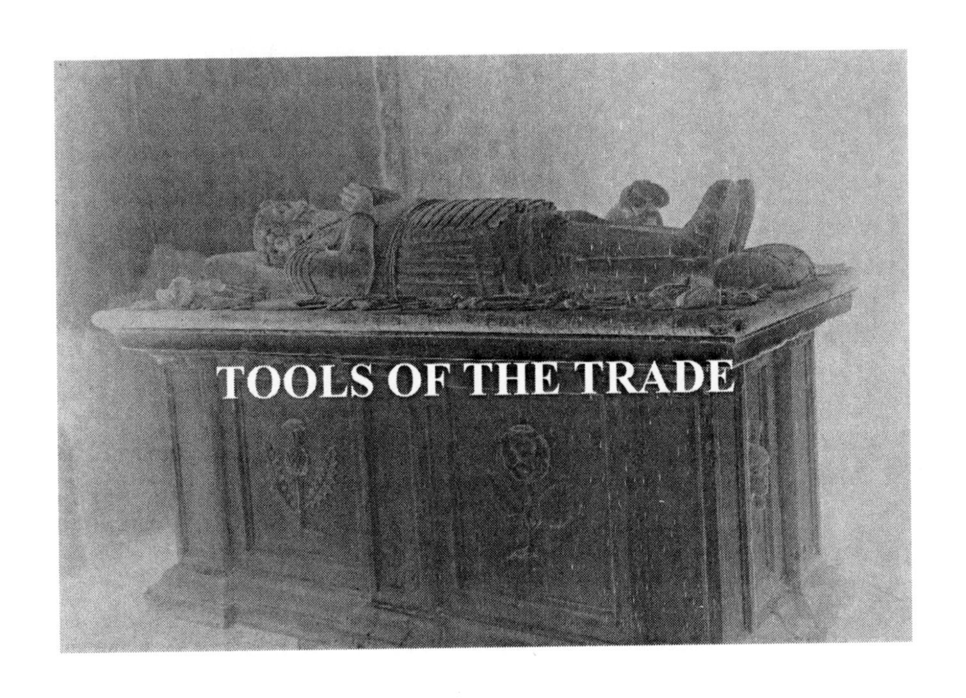

"Man is a tool using animal…nowhere do you find him without tools: without tools he is nothing, with tools he is all."

- Thomas Carlyle

If you are a ghost hunter then I would bet that you have some tools of the trade. Maybe it's an EMF meter, a digital recorder or perhaps a camcorder. You may have one or two items or an entire arsenal of high tech tools. But you *have* tools. I also have them and use them during investigations and you should, too. But using them correctly and understanding their role in a paranormal investigation is important. And tools are not a replacement or substitute for your senses as we will see in the next chapter.

What are the tools of the trade for ghost hunting? There are a lot of them. Some are common and some a bit out there. Perhaps the most common, and I would argue most important, tool is the EMF meter. EMF stands for electromagnetic field. A variety of meters, ranging in price, can be used to measure these fields.

A digital image of my K-II Meter in use during a recent investigation. The toggle switch on the side was installed to prevent false positive readings. The K-II meter has become the meter of choice for paranormal researchers.

An infrared digital image of me with an EMF meter on the main staircase of Boudlerberg Manor in Stony Point, New York.

EMF meters are the one tool that you should definitely include in your ghost hunting kit. It can be used to provide important information on what is going on around you during an investigation and alert you to potential paranormal activity.

The EMF meter is also easily abused. Used incorrectly, the EMF meter will provide results that are skewed and inaccurate. If you misinterpret these readings you may make conclusions that are simply wrong. As with any scientific instrument using it correctly is essential to obtaining accurate information.

Skeptics love to criticize paranormal researchers for using equipment to find something (ghosts) that it was not meant to find. They almost always illustrate their point with the EMF meter, proudly saying that ghost hunters use these meters to look for paranormal activity when these meters were never designed for such a purpose. This is where we can turn the tables on the skeptics and do a little skeptic bashing! The skeptics are wrong. Sensible paranormal investigators do not use the meters in a manner inconsistent with what they were designed to do. We use them to measure electromagnetic fields. That is what EMF meters are designed to do. If you call your EMF meter a ghost detector or ghost meter and walk around like a character out of the movie *Ghostbusters* then you are the person that the skeptics are referring to. The EMF meter is not a ghost meter. It is not a ghost detector. We do not use the EMF meter to find ghosts or measure ghosts. We use them to find and detect electromagnetic fields. If you call your EMF meter a ghost detector or tell people that you use it to find ghosts – please stop! You're hurting our credibility and doing a disservice to all ghost hunters who work hard to find solid evidence of hauntings.

When used correctly the EMF meter is an excellent tool for indirectly measuring paranormal activity. We are not necessarily detecting and measuring a ghost, but detecting and measuring the influence that ghosts may be having on the atmosphere around him or her. It is thought that as ghosts attempt to manifest they either generate electromagnetic fields or draw upon energy surrounding them which casues higher electromagnetic fields. Using an EMF meter we can look for inexplicable fluctuations in electromagnetic fields and use them as a guide in an investigation. For example, if you are using an EMF meter and start to pick up a higher than expected reading and at the same time record an EVP you may have better evidence of activity than if you had just measured the EMF or EVP alone.

Indirect measurement is frequently used in science and medicine. If you have ever gone to the doctor and had your blood pressure checked it was most likely done by an indirect method. The doctor or nurse put a blood pressure cuff on your arm, inflated it and listened to certain sounds through a stethoscope. That is how they determined your blood pressure. That is the indirect measurement of arterial blood pressure. The alternative direct method is invasive and requires an arterial line inserted directly into an artery. That certainly does not sound practical for your yearly check up.

I studied and worked in the field of exercise physiology for many years before going into law enforcement. We used indirect measurement all of the time. We palpated heart rate at the wrist and assessed oxygen consumption though various tests that gave us an indirect measurement. We assessed body fatness though skinfold measurements, another indirect method. All valid when using appropriate equipment, protocols, and correct interpretation.

The same is true for the EMF meter. The first thing you should do at any haunted location is determine the presence of electromagnetic fields. They occur naturally and also by manmade sources and many places have various levels of electromagnetic fields. Just because your meter indicates that there is some electromagnetic field fluctuation it does not mean there is concurrent paranormal activity. Especially if you find high, steady readings or readings that increase as you get closer to a wall or other potential source of the electromagnetic field. These are almost always going to clue you in to their manmade origins.

We have been on many an investigation where we picked up rather unusual readings only to find out that a refrigerator or other electrical source was to blame. Old homes can be especially problematic as the wiring is usually less than desirable and leaky outlets and unshielded wires lead to higher readings. It is easy to get excited on an investigation when you observe a higher than expected EMF reading. That excitement quickly fades when you locate the source it is manmade in nature.

EMF fluctuations can also be caused by radio interference, cell phones and even movement and the human body itself. Different meters have preset sensitivities and that will dictate what you can detect. One meter we use is the Trifield Natural EM Meter. It's a triaxial meter, meaning it measures the EMF all around the meter to determine a signal. The Trifield Natural EM Meter is so sensitive that it can detect humans. It is thus not a good hand held meter (there is a setting that allows you to hold it, but this is not a technical manual). I have seen people on television shows walking around with these meters in their hands and showing off the activity that they were picking up. Of course they were picking up activity – human activity! The Trifield Natural EM Meter has an audible alarm that can be set to make a subtle noise as it detects fluctuations in the electromagnetic fields. It zeros out the stable background EMF then picks ups fluctuations from that. We use them in strategic locations on our investigation and put them near a camcorder or audio recorder so that later when we listen to the recordings we can hear if there were any unusual EMFs.

Infrared digital image of the Trifield Natural EM Meter in use during an investigation at the abandoned Tamarack Lodge, Greenfield Park, New York. We were picking up unusual fluctuations outside one of the old buildings.

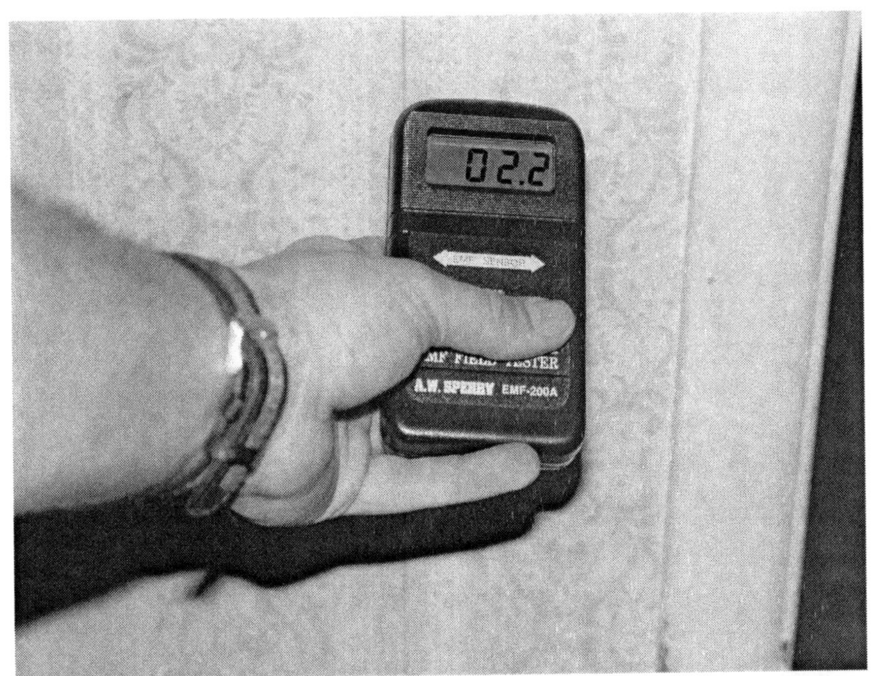

Digital image of my Sperry EMF meter in action. The reading is pretty high here because my refrigerator is on the opposite side of the wall. This is a handy meter but the digital display is difficult to see in the dark.

Another popular meter is the Cell Sensor EMF Meter. This is a single axis meter, meaning it measures whatever it is pointed at. The Cell Sensor has a probe which attaches to the unit and the unit lights up and makes a beeping sound when it detects electromagnetic fields. It is a handy tool and works well. The main problem is that the probe is separate from the meter unit, so if you lose the probe you have a meter that is only good at detecting cell phones! It is also a bit more cumbersome in that you have a meter unit to hold and also a probe to plug into it then hold and use to measure the area you are investigating. I have seen investigators put the meter unit in a pocket and just hold the probe, which is a good solution as it beeps and lights up when EMFs are present. This eliminates the need to watch a digital display. The flashing lights can cause problems for pictures and video and the beeping sound will all but ruin your chances of getting EVPs. I have only used the Cell Sensor EMF Meter once or twice on an investigation. It worked well but I just could not get used to the probe and wire so my Cell Sensors end up sitting in an equipment case most of the time.

Ghost hunting twins using their Cell Sensor EMF Meters. A digital audio recorder is on the large stone in the center.

If you like using the Cell Sensor and have good results with it by all means keep using it. It has a good range for detecting EMF fields and a does not detect fields at a great distance from the probe. This is great for reducing the likelihood that your EMF readings are caused by artificial sources rather than supernatural ones! But here is a word of caution: moving the probe too quickly can cause the meter to alarm, giving you a false positive reading. When my sons want to go ghost hunting (and yes they do want to do that) I let them use the Cell Sensors because they get results. Maybe not reliable ones, but then of course, they are only kids and they enjoy going ghost hunting with Dad!

When my kids run around the cemetery excited over activity on their Cell Sensor EMF Meters they don't quite understand that they are picking up movement of the probe and not ghosts. When they do it is cute. When grown adults who are paranormal investigators do it they are just unprofessional and sloppy. Know how your meter works and know what can cause it to give a false positive reading.

Probably the meter growing the fastest in popularity is the K-II Meter. This meter uses a series of LED lights to indicate the presence of electromagnetic fields and does not make any sounds. It is a great hand-held meter because you can see the LEDs easily in low light conditions (ever try reading a digital display in the dark?). Many groups, including Linda and I, have had great results using this meter on investigations. We have had considerable success with using it as a tool to facilitate communication with unseen entities. We set the K-II meter in the room and explain to any spirits present that if they go near the green light (that is the baseline light on the K-II) that they can make the lights change color. This has worked for us so try it out on your next investigation. Stay away from technical language (if a person in life had no clue what an EMF meter was then they most likely have no clue in death as a ghost!) and use simple, clear instructions.

One problem with the K-II Meter is that to turn the meter on you have to push on the power button with your thumb and then hold it down. If you lighten the pressure a bit you can inadvertently turn the meter off. A little more pressure can then turn it back on. When this meter turns on all of the LEDs will light up and blink as a self test. If you don't know that you have accidentally turned the unit off and on again you could misinterpret this self test as being paranormal activity. A low cost solution to this is to wedge a coin into the on button to keep it on during an investigation. A more complicated approach is to have a toggle switch hardwired into the unit. This toggle switch takes the place of the on-off button and eliminates any potential false readings due to the built in on button. There are sites on-line with the electrical schematics for this remedy. Thankfully, Linda's husband Bob is great at such projects and he was able to read the schematics and wire in toggle switches. The K-II is a surprisingly strong meter and can pick up EMF fields from man made sources such as cell phones. Your cell phone does not have to be in use in order for it to generate a signal that can potentially contaminate your evidence. I have tested the K-II meter and my cell phone sets it off from across the room when receiving a text or even when a call is about to come in. The best way to eliminate this problem is to leave your cell phones off on an investigation and keep them in a neutral area where you are not looking for EMFs.

Regardless of the meter you use make sure that you use it correctly. Look for manmade sources of EMFs at any haunted site you are investigating. When you do come across an EMF anomaly try and rule out potential sources around you. Look around and often you will find a logical explanation. When you don't find the cause consider what other evidence you are encountering at that moment. Are you feeling anything unusual? Is anyone feeling something? Is there a cold spot? Do you hear something? These are the experiences you hope to validate from EMF fluctuations that have no earthly explanation.

The EMF meter is by no means the only tool that ghost hunters use. Thermometers are another, and generally the infrared non-contact thermometers are common. These are ones that you hold, point, and get a temperature reading to show up on a digital display. These are great tools but require the infrared beam to bounce off of something in order to measure the temperature. Thus, you are measuring surface temperature versus air temperature. A better alternative for measuring air temperature (such as when you are feeling a cold spot) is a probe thermometer. A small probe can be held into the cold spot and then a digital display will read the temperature. You can measure the temperature of surrounding air to confirm the presence of a cold spot. A cold spot is not necessarily indicative of a ghost but again, when considered with the totality of the investigation, may help you draw your conclusions.

Many researchers rely too heavily on the equipment used in paranormal research. Yes, cameras, audio recorders, thermometers, and EMF meters are valuable assets on an investigation. You can use them to validate experiences and obtain very credible and reliable evidence of paranormal activity. Far too often researchers become too focused on the equipment and lose site of the personal involvement that is unavoidable on investigations. As my good friend and ghost hunting partner Linda Zimmermann always says: sometimes the best way to investigate is to sit quietly and just listen. The next chapter will focus more on the role of personal experiences as valid evidence of the paranormal.

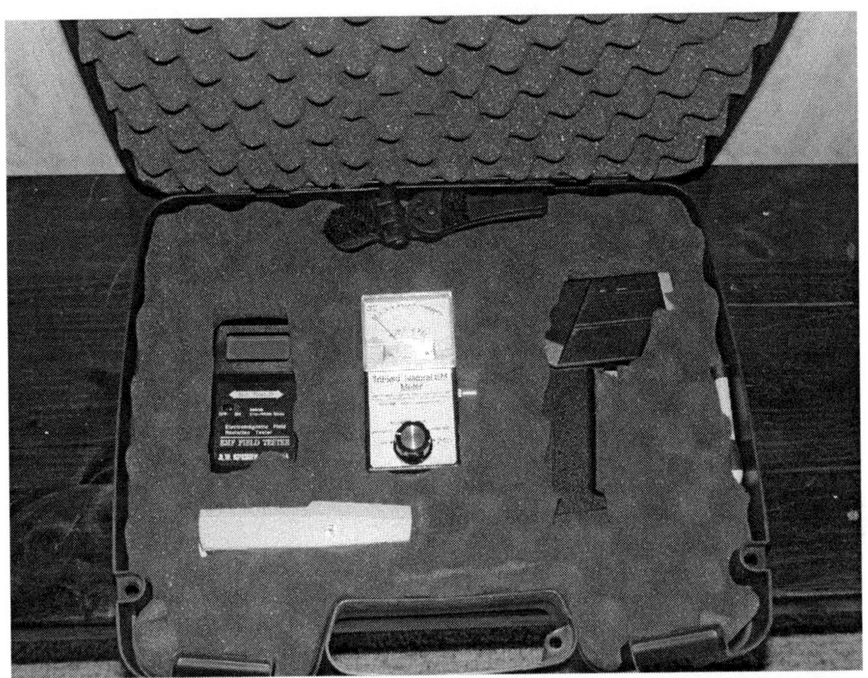

Protect your investment. This picture shows how I have placed my EMF meters and non-contact thermometer into a protective case. I buy the hard plastic sporting equipment cases that are available at any sporting goods store or in the sporting good sections of most major retail stores. The interior is foam that can be cut to custom fit your equipment. Storing and transporting your equipment in cases such as this is an inexpensive way of protecting your equipment so that you get years of ghost hunting enjoyment from them!

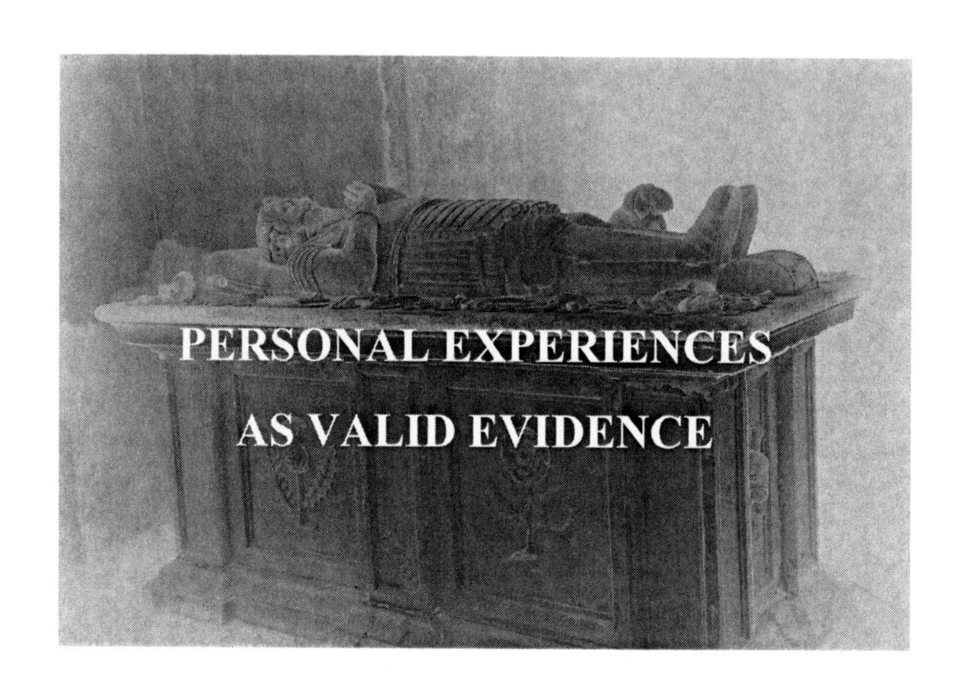

"You need not leave your room. Remain sitting at your table and listen. You need not even listen, simply wait. You need not even wait; just learn to become quiet, and still, and solitary. The world will freely offer itself to you to be unmasked. It has no choice; it will roll in ecstasy at your feet."

– Franz Kafka

You are a ghost hunter. You have gathered up your gear and charged up the batteries. You and your crew make the short trip to a haunted house near by. You set up your cameras and recorders and other gadgets. Then you put in earplugs, blindfold your eyes, put on a nose plug and...sound about right?

The scenario above is a bit silly. No self respecting paranormal investigator would really go to a haunted location and blindfold themselves, or plug their ears, or otherwise limit their senses. But many investigators do just that. Not in an obvious way as with blindfolds and ear plugs, but by discounting and disregarding personal experiences and encounters. If it can't be recorded or photographed then somehow it isn't valid evidence.

Is this good ghost hunting practice? Is a personal experience valid evidence of paranormal activity? Based upon my experiences at numerous haunted locations over the past eight years, I believe that they are.

First, consider how most paranormal investigators learn of a haunted residence: from people that have experienced paranormal activity. Common experiences include:

> "You're going to think I am crazy, but I see the form of an old lady walking along my hallway at night."
>
> "I live in an old farmhouse where the previous owner shot himself in 1987. In one certain spot I smell the odor of gunpowder and pipe tobacco. He used to smoke pipes and I recently learned that this is the spot where he killed himself."
>
> "I hear footsteps on the stairs and someone walking in the attic at night. I check and no one is there."
>
> "Whenever I sit at my desk to pay my bills I feel as if someone is standing behind me. I spin around, but no one is ever there."
>
> "We live in an old house where in 1850, three people died of small pox. At night when I am sleeping I am often awakened by the sensation of someone lightly stroking my

arm. Turns out, a mother and her two children died of the disease here…the mother the last to go."

This is what people tend to describe when reporting their alleged hauntings. They tell about things that they see, hear, or smell. They talk of feeling things, or even of being touched. They describe sensory experiences. What is uncommon – no unheard of – is to hear someone call and tell you that they have unusual EMF fluctuations and have measured temperature differences.

Think of some of the great ghost stories from literature. What types of scenes do they portray?

> "I muttered, knocking my knuckles through the glass, and stretching an arm out to seize the importunate branch; instead of which, my fingers closed on the fingers of a little, ice-cold hand! The intense horror of nightmare came over me: I tried to draw back my arm, but the hand clung to it, and a most melancholy voice sobbed, 'Let me in—let me in!' 'Who are you?' I asked, struggling, meanwhile, to disengage myself. 'Catherine Linton,' it replied…"
>
> Lockwood's encounter with the ghost of Catherine
> Emily Brontë's *Wuthering Heights*
> Excerpt from Chapter 3

The classic encounter with the ghost of Catherine in *Wuthering Heights* is one of literature's most chilling scenes. Brontë describes the desperate ghost of Catherine and how she touches Lockwood's hand while calling out to him. *Wuthering Heights* was published in 1847, long before EMF meters, video cameras, and thermal temperature guns. There is no reference to modern ghost hunting techniques. But there is mention of a personal encounter with a ghost. After all, isn't that what a haunting is about? And how exciting would that same scene from Wuthering Heights be if we excluded the personal experience?

> "I muttered, knocking my knuckles through the glass, and stretching an arm out to see if I could detect any EMF fluctuations on my K-II Meter! The intense horror of

nightmare came over me: my EMF meter was indeed picking up a higher EMF field, and a most melancholy voice sobbed, but since I did not capture it on my digital audio recorder, I can not use it as evidence."

Lockwood's encounter with the ghost of Catherine as modified by the author to make his point!

That may be a description appropriate for a book detailing paranormal investigations, but it would make for lousy classic literature. And if, as an investigator, that really did happen to you, would you really want to disregard it? Would you really leave that location, tell the story, then shake your head and say that it doesn't really count as evidence?

The best stories, from classic literature to the scariest of movies, portray encounters with spirits as involving the senses. They are seen, heard, smelled or felt. Phantom footsteps on the staircase are certainly scarier than fluctuating EMF fields on the staircase!

Hauntings, by their very nature, are sensory experiences. *We do not measure hauntings, we experience them.* Ghosts are, after all, remnants of once living people. It is common sense to expect that ghosts are going to be detected by the same sounds, smells, and feelings that one gets from living people. If you have a family member who smokes pipe tobacco, you will smell him before he enters the room. You will hear family members walking in your house – going up or down the stairs, walking on the floors above you, or closing a door. You will hear them talking in adjoining rooms. These are normal encounters with the living. And since ghosts were once living these are normal things to experience from them as well.

Have you ever been around someone who is very angry? Or very mean? Or very depressed? Have you ever observed how someone who has a very powerful personality can walk into a room and you can feel it? I would bet that we all have experienced this. We can often pick up on other people and their feelings. A negative person often just feels that way. You know it, you feel it, and you don't want to be around them. Some people are so happy and easy going that we are drawn to be around them. Sometimes they are too happy and you just want to slap them back into reality! When we are around someone happy and content they feel good and in turn we feel good around them.

Ghosts were once people. They had a personality. Imagine the broad types of personalities you would get by sampling a random group of people. No two people would be exactly alike. Even my identical twin sons have different personalities. So, too, will you find this with ghosts. You will find happy, sad, angry, confused, and any number of other traits amongst the spirits you encounter. I believe that most ghosts are in death, what they were in life. An angry person who dies and does not move on will most likely be an angry ghost. A happy person in life may be a happy person in death if he or she were to be a ghost. A shy person who avoided people may become a shy ghost who is elusive, whereas a social butterfly in life, may be the life of the party in death. If ghosts are us, our essence, our personality, then we continue on as we were. So feeling those personalities is not something that should be avoided or cast off as irrelevant.

Feelings may also be more than just the personalities of the ghosts. On many investigations we have felt cold spots. It is theorized that ghosts, as they attempt to materialize, draw energy from the air causing the temperature to drop. Cold spots are a commonly reported phenomenon in haunted locations. Often they can be documented using some type of thermal scanner or thermometer. Other times, they happen quickly during an investigation and are fleeting. When there is no explanation – drafty windows, duct work, etc, then they may be clues and evidence of the paranormal activity at a location. Would you simply discount this?

While investigating a building in Ellenville, NY, Linda and I had been in an attic location that was accessible by a long, narrow staircase. We had gone up and felt quite comfortable in the attic. It was the staircase that frightened us and the noises that we were now hearing from the area at the bottom of the stairs on the floor below. Linda made a comment that we had to go down the stairs and neither of us felt that it was a good idea. However, there was only one way down and we had to go. Linda suggested holding the railing tight and headed down first. I followed behind with my light. Towards the bottom Linda observed a shadowy figure about thirty feet away rush out of the door. We attempted to recreate this shadow using my flashlight and our positions on the stairs. We could not. Linda had seen something that we could not explain. We did not discount this incident nor disregard it as evidence.

Still not convinced about the case for personal experiences? How many of you reading this have ever received some form of a traffic ticket? If you have ever received a ticket and then later contested it in court, chances are the only evidence that would be presented would be the testimony of the police officer. Testimony about what he or she had seen: you ran the red light, or didn't stop at the stop sign, or were speeding. The officer will testify to what he or she had observed that constituted a traffic offense. Even when it comes to the use of police traffic radar, the testimony of the officer observing the speed indicated on the radar unit is sufficient testimony.

Let me draw upon some more police analogies. A police officer responds to a report of a large fight. He arrives to find a group of people surrounding someone lying on the ground. That person is dead with a large knife sticking out of his chest. Everyone in the group points to the same person and says "that's the guy who stabbed him." You shrug your shoulders and tell them "unless I have it on video then I have to discount that as evidence." STOP RIGHT THERE! If this were to really happen, that cop should lose his job. Witnesses to the crime are giving crucial identification evidence and that cop should have taken action and arrested the bad guy. The detectives can process the scene and gather forensic evidence further proving the case. But that initial eyewitness identification is crucial and would be certainly a big part of any criminal case.

What people see, hear, or even smell at a crime scene is often crucial evidence of a crime. An investigator at the scene of a suspicious fire would not discount the odor of gasoline. An unsolved shooting may involve witnesses who heard gun shots to help determine the time of the crime and potential location of the shooter. Those same observations are equally as important to paranormal investigators, as both evidence and a guide to focusing our investigation.

As investigators we come to a residence or other location and spend a limited amount of time there. In many cases these places are quite large, with multiple rooms and hallways and other spaces. It is almost impossible to cover every square inch of a place during an investigation. How, then, do you cover a haunted location adequately? You talk to the people that have witnessed or experienced paranormal phenomena. You need to hear from the people who live, work or otherwise frequent those locations. Then, you use this information to focus the investigation and validate your own experiences, if any.

When we investigate a haunted location, we always ask what people experience. We want to try and grasp the nature of the haunting. What are people experiencing? Where are people experiencing it? When are people experiencing it? Few of us have unlimited resources, or the unlimited access to investigate every square inch of a residence or other haunted location. So we try and focus our investigation to those areas where activity has been noted. It's kind of like fishing in a large lake. Would you just go out in a boat and start casting your line randomly, hoping to find fish? You could and you may catch something, or return empty handed. Anyone who has ever gone fishing will tell you to go where the fish are – where other people have noted catching them, or where you know that fish are often found. You may even use an electronic device to find fish under water. Going where the fish are? Electronic devices to find fish? Starting to sound like ghost hunting now, isn't it?

In many cases you will find a long history of experiences and encounters at a particular place. My own grandparent's home in Port Jervis has several decades- worth of experiences that span several generations. When Linda and I began our first investigation there we did not discount these personal stories. We collected them, discussed them and focused our investigation – and subsequent follow ups – based upon those experiences as well as our own.

As an investigator you will also become more sensitive and tuned into the phenomena that are going on around you. It doesn't mean that more is happening because of you, but may mean that you are simply more aware of it. I give you yet another police analogy (there are going to be a lot of them!): when I was a rookie on the street learning the ropes I found out how quickly and easily it is to spot vehicle infractions. I would be riding down the street off-duty with my girlfriend and point out expired inspection stickers, missing front plates, cracked windshields, no tail lamps, and no seatbelt use. And then there were the moving violations: the speeders, the red light runners, and the people that rolled through stop signs. It became second nature. It was not as if more people were committing violations in front of me, just that I was more aware of those violations. Spotting them became second nature.

It is this way with ghost investigating. You will develop your ability to feel what is going on around you. You will – or should – become more sensitive. This is a good skill to develop as ghost hunter.

Learn to listen to those perceptions and feelings on investigations. Not only may you discover fascinating evidence and a new dimension to the haunting, but it may be an internal warning system. Sometimes knowing when to leave – or feeling when to leave – is as critical as knowing where to set up your equipment. Linda and I both have developed this sense over the course of our investigations and we never question one another when a red flag goes up and one of us feels it is time to go or we are feeling something uncomfortable or threatening.

I had a personal experience at the former Ulster County Jail in Kingston, New York. During our investigation I had decided to sit at a metal table within a cell block. I was sitting with my K-II meter and listening, hoping to hear something unusual. Suddenly, I had an overwhelming sensation that I was choking. I felt as if I could not breathe and the sensation grew in such intensity that I needed to get up and walk away from the cell block. Weeks later I learned that a man had hanged himself inside the cell near where I was sitting. Is this not valid evidence of paranormal activity?

You may also find that personal experiences are accompanied by validation through instrument such as EMF meters. In these instances not only are you experiencing the haunting on a subjective and personal level, but you are also measuring it or documenting it objectively with scientific instruments. You may, for example, feel a cold spot and be able to measure that cold spot using a thermometer. You may hear foot steps and record those footsteps onto a digital audio medium. You may even feel as if something is touching you, or standing next to you, and at the same time you measure a high electromagnetic field. When a personal experience is documented you have compelling evidence of the paranormal.

Former cell block at the abandoned Ulster County Jail, Kingston, New York. Notice the picnic style tables in the block. This is where I experienced a choking sensation outside a cell where a man had hanged himself.

A very personal experience at the Iron Island Museum in Buffalo, New York, this past summer, is among the most moving that I have encountered. It is also a great example of a personal experience validated by scientific instruments. Iron Island Museum is housed in a building that began as a Methodist church in 1885. In 1965 the building found new life as a funeral home. The funeral home business closed in the 1990's and in August 2000 a museum opened there to celebrate the history of the area, the military and the railroad.

We were sitting in one of the former viewing rooms with our hosts and I had been saying aloud to the spirits that we had driven some distance and really wanted to talk with them. Linda mentioned that she thought she saw some movement coming from a back corner of the room and one of our hosts, Jennifer, asked if it were like a child peeking. Linda agreed that it was and it was explained to us that this is one of the common apparitions in the museum – a ghost of a child they called Tommy.

I was seated with my back towards the direction of the figure and had been holding my K-II EMF meter with my left hand. I slowly moved my hand off to the side and slightly behind me, pointing it in the direction of the figure that Linda had seen. I began to explain in a calm, soft voice not to be afraid and that if he wanted he could come close to me and play a little game. I gently explained, in the simplest terms to an unseen child, that if he snuck up next to me he could make the lights on my toy change colors. Within a second my hand was ice cold! Simultaneously, the lights on the K-II meter lit up indicating the presence of an electromagnetic field. This electromagnetic field had not been there moments earlier and was coinciding with this very cold feeling on my hand.

As my K-II meter reacted and my hand felt cold I had the distinct impression that this child was *holding my hand.* An incredible sadness and loneliness came over me and I felt tears begin to well in my eyes and roll down my face. I maintained my composure and thanked the boy, telling him he did a great job. I asked the little boy if he liked trains and Jennifer recorded a sound on digital audio that sounded like a boy giggling. I also asked him if he could tell us his name and we later heard on digital audio the voice of a little boy clearly say 'no'.

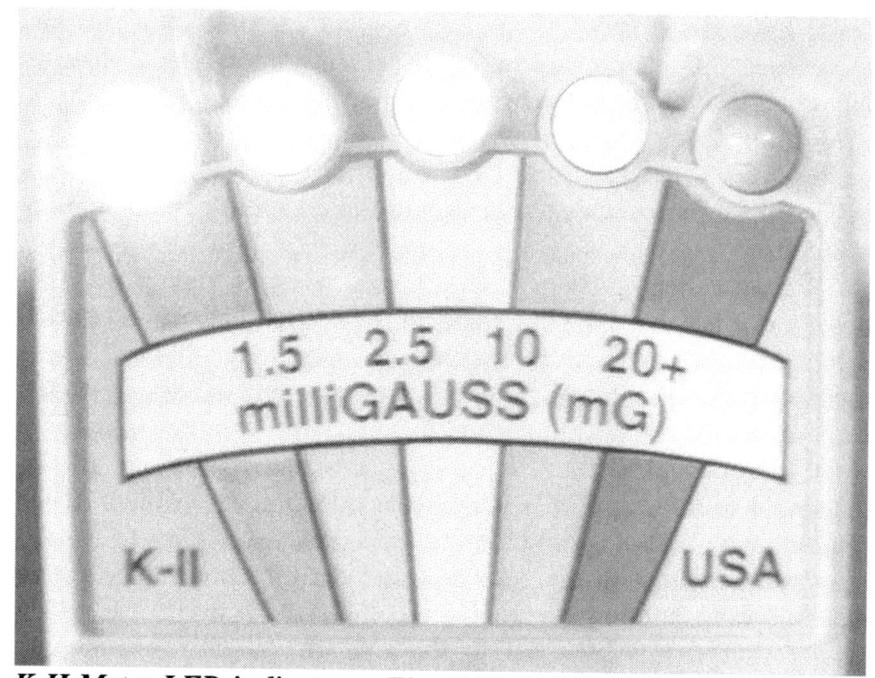

K-II Meter LED indicators. The K-II reacted and measured fields in the range of 2.5 mG and higher during my encounter with the child.

My interaction with the phantom child did not last long as a sudden shift in energy came over the room and as quickly as the child's presence came, it went. The lights on the K-II meter went still and warmth returned to my hand. I needed a break and we left the room to regroup in a kitchen area where we had stored our equipment. I was still feeling overwhelmed with emotion - thoughts and feelings were fresh on my mind and I needed to write them down. And I needed to do it quickly. I asked Linda for her notepad and quickly scribbled down notes. My handwriting was fast and full of emotion. Amongst the feelings that I noted was the overwhelming feeling of loneliness and sadness. I also wrote 'touching my hand' and underlined it, emphasizing the very intimate nature of this experience.

What fascinates me to this day is that I also wrote down how I felt that the little boy was sorry and he just wanted to play but should know better. It was my distinct impression that another male entity had come into the room and *made* the child leave. That is why the child was sorry and thought that he should know better. Whoever the angry male entity was, he clearly was a bully.

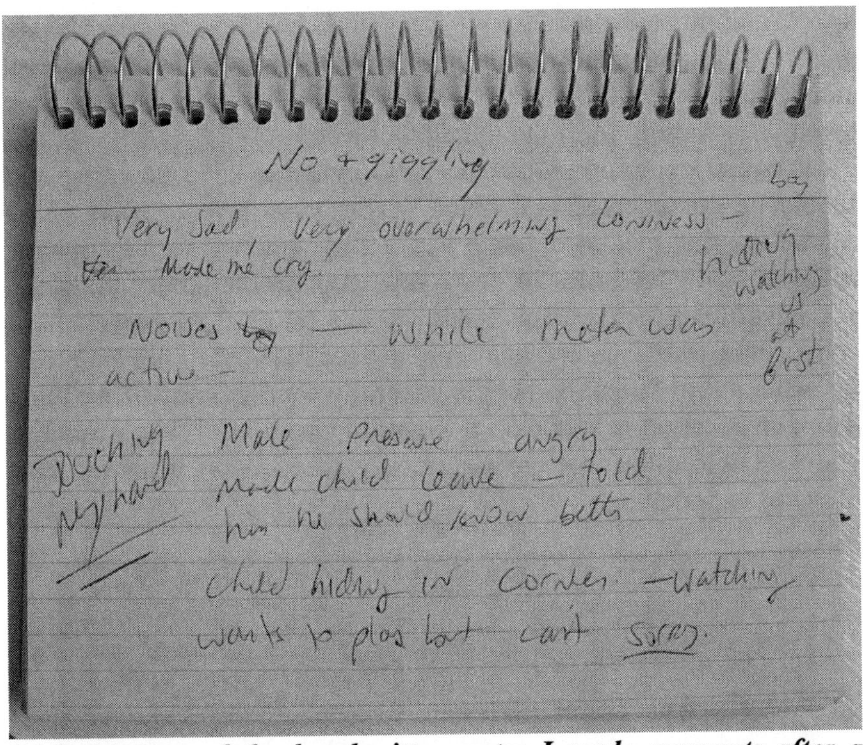

Digital image of the handwritten notes I made moments after a touching and moving experience with the ghost of a little boy. Photo courtesy Linda Zimmermann.

My encounter with the little boy in the Iron Island Museum was hard to forget and I admit that I was a little less into the investigation after that. We did take some time to verbally chastise the paranormal bully (and it worked – read the full investigation in *Ghost Investigator Volume 9: Back from the Dead* by Linda Zimmermann) but emotionally I was drained. It is hard to go into an investigation and turn off your emotions. I had connected with the lost, lonely spirit of a little boy; a boy who I felt was probably not much younger than my own boys who were five years old at the time. As a father my heart ached at the thought of this poor soul. Even now, writing this, I am moved to tears thinking of that child. I don't know his name, I don't know why he is stuck there, but I know that for a brief moment the veil between our worlds was transcended and the spirit of a lost, lonely child reached out and held my hand. Perhaps for a brief moment he felt the love of his own father and a sense of safety from the dark, overbearing male presence that eventually made him leave the room. I wish I knew more about him and I hope that someday he finds his way to those loved ones waiting for him on the other side.

Personal experiences can be valid evidence of a haunting and should be documented and noted during your investigation. You should use that evidence, along with other evidence such as photos, audio or EMF readings, to draw your conclusions about the place you have investigated. This leads me to a bit of a pet peeve I have about some ghost hunters; declaring a place haunted or not haunted.

After an investigation people generally want to know if we think that their location is haunted. It's a valid question. I would want to know. Quite often that answer is easy based upon what we have found or experienced.

The author descending the main staircase at the Napanoch, New York Shanley Hotel. Photo courtesy Linda Zimmerman.

Our first investigation at the Shanley Hotel in Napanoch, New York, was one of those occasions. It was such an onslaught of experiences that Linda and I both left the hotel that evening physically and mentally drained. We felt as if we had been sucked up into a paranormal tornado. There was no need to ponder hours of evidence in that case. We had lived it. We had experienced it. There was little doubt that the Shanley Hotel was haunted.

Then there is the opposite example of the Patchett House in Montgomery, New York. The Patchett House is the current home of the Wallkill River School and Gallery. It is meticulously restored and a wonderful space. In previous incarnations, the Patchett House served as a private residence, boarding house, and a funeral home.

Our first investigation at the Patchett House was rather uneventful. I had made a comment to a reporter that was present that just because we really did not have anything going on did not mean the place was not haunted.

Infrared digital image of the embalming sinks in the basement of the Patchett House. A Trifield meter is visible above the center sink.

We had spent a few hours in a place with over a hundred plus years of history behind it. Credible people had witnessed and experienced paranormal activity. Just because we didn't walk away with the supernatural smoking gun did not mean that the Patchett House was not haunted. It meant that at that time we did not have any experiences or scientific evidence to document that haunting.

Several months later we returned to the Patchett House for a follow-up investigation. This time, things were different–*very* different. We had numerous experiences, including unexplainable crashing sounds, unusual EMF fluctuations and even captured two pictures of some type of apparition. Had we drawn our conclusions from the first investigation we would have erroneously labeled the Patchett House as being 'not haunted'. Being patient and returning a second time was a wise choice and shows that in this field, you can never base your conclusions on one quiet, uneventful investigation. We are also fortunate in that we tend to research locations in advance, a good idea for any paranormal investigator or group. Talk to the people involved in the location and ask what is being reported. Listen to the history of the place. Ask how long the activity has been seen or heard or experienced. Ask how many people have experienced it. Get as much information on the location as you can. Ask if there is any particular significance to the site (e.g. a murder, suicide, tragedy).

Choose locations with a history of haunted activity, a significant history in general and credible witnesses to the paranormal activity. You will increase the likelihood that you are going to find activity. When I was a young street cop and wanted to write speeding tickets, I grabbed the radar unit and went out to the roads that were more heavily traveled and people had complained about speeding. I did not go sit on a dead end road that had one car per hour travel down it. I went to the location where it was more likely to find speeders. It is the same with ghosts. Go where you are likely to find them. And on those occasions when you go and find nothing – and it will happen – do not be quick to jump to conclusions and declare the place not haunted.

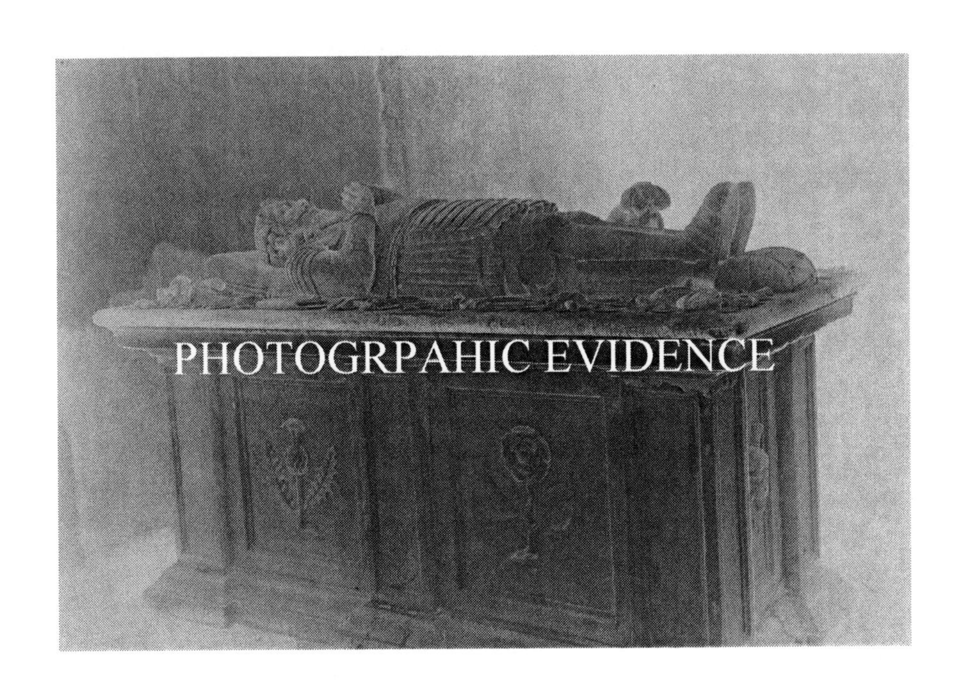

“A photograph is usually looked at - seldom looked into.”

– Ansel Adams

It is said that a picture is worth a thousand words, and a picture of a spirit or ghost is certainly worth that and more. For most paranormal investigators, a photo of an actual ghost – the apparition of a deceased human being – is *the* Holy Grail of evidence. Unfortunately, these full bodied apparitions are rare. One of the best apparitions was captured by Linda Zimmermann during an investigation a Pomona, New York. That photo is reprinted in this chapter. The complete story can be found in *Ghost Investigator Volume 6: Dark Shadows*.

It is more common to see other types of paranormal phenomena passed off as evidence of ghosts. Conduct a search of 'ghost photograph' on a search engine of your choice and you will likely be able to spend an entire afternoon – or longer – pouring through sites loaded with ghostly photos.

The camera is a basic tool for paranormal investigators and we use both video and still digital mediums. Much of what you are going to read in this section applies to video as well as still images. Essentially, both can offer wonderful mediums for capturing a ghost or apparition. Both are also problematic in that you can easily obtain eerie images that appear to be paranormal when they are simply the result of common errors or natural conditions. So the advice in this section may be focused on still photography, but apply the information equally to obtaining good video evidence as well. In fact, video itself may be especially problematic in that it is dynamic – it is constant. Conditions may change from one moment to the next. Passing vehicles, sounds, lights, shadows, flashlights, infrared illuminators can all create unusual-looking phenomena in recorded video. An infrared illuminator reflected off a mirror may not be visible to the unaided eye but may create supernatural-looking anomalies on the recorded medium.

We use our camcorders – Sony Handicams with Nightshot infrared mode – mostly as stationary camcorders set up at key locations. We then set them to record and leave them alone. If we are going to be in the area where we have a camcorder set up we generally announce it loudly enough to be recorded on the video so that later, during playback, you know that potential shadows, sounds, and lights may be attributed to the investigators being in the vicinity of the camcorder.

This is important. We spend hours at a location and often have hours upon hours of video to sit through. Have you ever stared at a monitor playing an infrared video from a room at a haunted house? How about for two hours worth of video or more? Now think about having three or four camcorders, or even more, in a large location. Your mind plays tricks on you and you will go batty after awhile. You will also not remember every moment of the investigation and where you and every member of your team was at the time something was recorded. So use your voice and announce when you are nearby locations – especially if a camcorder is in the vicinity. It doesn't hurt. It gives you a real-time accounting of who is where and what is going on so that you do not think the weird light popping up is a ghost, when in fact it is someone walking around with a flashlight. There has been debate amongst investigators about using film versus digital medium. When I was a new investigator I used a 35mm camera and would have to load it up with film and, of course, have the film later processed and pictures printed. It's time-consuming and the results are not quite as instant as one would like. Sure I have an old Polaroid instant camera – but the cost of film made it impractical to use to any great extent.

I am not going to debate the advantages of film versus digital, nor am I going to go into the technical aspects of digital files and compression. For those interested there are plenty of resources to find out more about those topics. For us, digital works and works well. We take a lot of pictures on an investigation without any worry of development costs. The results are immediate and you can see, not only potential paranormal anomalies, but also those with more earthly explanations that you can make note of.

The biggest problem with digital is that it can be manipulated very easily using any number of commercially available photo-editing software. Fraud is thus a possibility. In my opinion, this is rare. More common, however, is the misinterpretation of photographic anomalies without taking the time to examine them with a critical eye. In other words, it is more likely to see good photos with bad interpretations. Not everything that looks paranormal is paranormal in nature. As investigators, it is critical that we take the time to carefully analyze photographic evidence. The extra time saves you the embarrassment of making a false claim and preserves your credibility as an investigator.

The following pages are going to address some of the common photographic anomalies that investigators encounter in the field. I have included many photos with apparent anomalies and their explanations in an attempt to demonstrate the importance of critical analysis.

We all want to find the Holy Grail of evidence. We all want to have a great picture of a ghost, apparition or some other form of paranormal manifestation. That is why we do what we do: to seek and find proof of the existence of ghosts. By taking the time to analyze, evaluate, and verify your photographic evidence, you not only enhance your credibility, but the credibility of the field of paranormal research.

And there will be some pictures in this section that I simply can not explain away….

ORBS

The author with his sister, Melissa, in the background at a private residence. 35mm film photo.

Many people, paranormal investigators included, believe orbs are proof of paranormal activity. Orbs, however, are problematic as evidence since they are easily produced by dust, moisture, water droplets, insects and other rather earthly causes. The photograph above, demonstrates how cameras with lenses in close proximity to the flash can create orb-like phenomena. This phenomenon does not discriminate between digital or print mediums. Cameras with this lens-flash arrangement create conditions that increase the likelihood that airborne particles will be illuminated and captured as glowing balls of light. And no, orbs do not have smiley faces in them. Save those for your instant messages.

We no longer use cameras in our investigations with this lens arrangement. Since this transition we find that we do not capture orb pictures on a regular basis. Orb pictures now are a rarity and easily explained away.

Outdoor photo taken with 35mm film camera.

The picture above was taken at the site of the Battle of Minisink, Minisink Ford, New York. Here on July 22, 1779, a hastily assembled militia met Joseph Brandt, a Mohawk Chief and Captain in the British Army and his army of about 90 Tories and Iroquois Indians. Brandt and his men had burned settlements along the Delaware River at present day Port Jervis, New York. Brandt was able to out flank the militia and claim victory. It is thought that 45 to 50 of the militia were killed and their bodies lay unclaimed, at the battle site, for 43 years until an expedition was led to recover the bodies and provide them with a proper burial. The orbs in this photo are attributed to dust or other airborne particles. Also problematic with this photo – as in any photo taken outdoors – is the likelihood that you are photographing insects.

The author pictured at Hospital Rock within the Minisink Battle Site. He is using an EMF meter to scan for any unusual electromagnetic fields.

Of interesting historical note is Hospital Rock, where Lt. Col. Benjamin Tusten, a physician, and seventeen of the wounded militiamen under his care were trapped and subsequently killed by the raiders.

Battlefields are considered 'hot spots' of paranormal activity. They are the sites of death, suffering, anger, hatred, misery and pure human emotion that are the stuff hauntings are made of. Be critical of photos taken outdoors at battlefields, cemeteries, and similar locations, as airborne particles may make more appearances than any paranormal visitors.

Death row. Eastern State Penitentiary.

Hospital Ward. Eastern State Penitentiary.

The preceding photos were taken as part of our December 2001 investigation at Eastern State Penitentiary, Philadelphia, PA. We were fortunate to be amongst the first investigators to gain access to the former prison and had nearly unlimited access to the vast complex.

The prison was in an extreme state of disrepair, dirty, dusty, damp, and moldy. Not surprisingly, we captured a lot of photographs with orbs in them. Both of the preceding photos were taken with a 35mm film camera. And yes, the lens-flash arrangement was close.

The photo below is an example of orb-like phenomenon caused by lens glare. This photo, taken with a digital camera, clearly shows how sunlight can reflect and create objects that look surprisingly paranormal in nature. Also keep in mind that cemeteries are problematic, as polished stone will also reflect light and cause any number of anomalies to be photographed.

Sun glare at a cemetery near Port Jervis, New York.

The photograph on the next page was not taken as part of a paranormal investigation. I took the photo in September 2002 as part of a nearly three week trip to Scandinavia. The photo was taken using my handy 35mm camera (the one that seems to attract orbs) inside Turku Cathedral, Turku, Finland.

Centered in the picture in front of the painting is a rather large orb. I note that this is the only orb picture out of 26 rolls of 35mm film that I took over the course of vacation. (Not to get side tracked, but do you have any idea how much that cost to process? Digital is simply more practical!). Also, I took quite a few pictures inside the cathedral and this is the only picture with any type of orb in it.

Now I am not sure what I captured in this photo and would love to be able to proclaim that I have photographed some type of sacred or holy image. I can not do that, however. My interpretation is that light streaming in from the large windows on either side of the orb contributed to the creation of this phenomenon. A dust or moisture particle close to the lens may also have caught the light from the flash and just randomly ended up so perfectly centered. I

I include this picture in presentations (and in this book) because the centering of the orb seems just too perfect. So I honestly have to say, I just don't know for sure. I will keep an open mind and allow you to decide for yourself what you believe is in this picture.

Photograph from 2002 taken inside of Turun tuomiokirkko (Turku Cathedral), Turku, Finland. 35mm film.

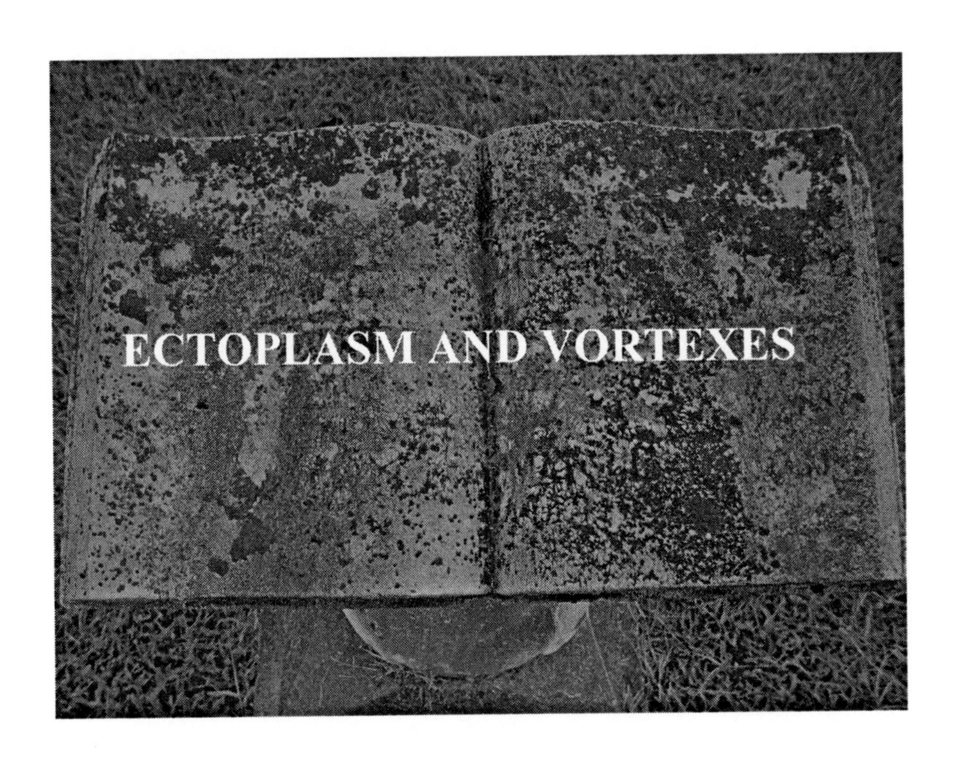
ECTOPLASM AND VORTEXES

Outdoor ectoplasm captured with a Sony F717 digital camera.

A second later the ectoplasm is moving away, nearly out of frame.

These photos were taken near a tree outside of Napanoch, New York, where an elderly woman hanged herself in the 1980s. The ectoplasm seems to be moving as I snapped several pictures in front of the tree. However, there is a slight problem with this. The pictures are bogus. They were staged.

Now before anyone questions my integrity, there was a good reason to stage these photos--to illustrate a fundamental rule of ghost hunting: NO SMOKING. If you smoke, you should quit and put the money you save by being smoke-free into some good ghost hunting gadgets. But if you do smoke, please don't do it on an investigation. These two pictures show how easy it is for smoke to be misinterpreted as paranormal activity. And I cannot blame just cigarette smoke: car exhaust, fog, and condensation from your own breath can all create similar anomalies.

Now I purposely staged these photos. While taking pictures of the tree (where an elderly woman *really did* hang herself) I noticed the property-owner smoking. As I was in the process of writing this book, I asked her to stand near me and smoke while I took a few pictures. When I got home and went through the photos I knew what to expect and knew exactly which set of images were taken while she was smoking.

But what if I had not set out to take these pictures? What if I was focused on taking pictures of the hanging tree and not paying attention to someone smoking nearby? I may have ended up with photos that I would think are paranormal in nature when they are not.

The lesson here is to *not smoke anywhere* during an investigation. No matter how far away you think you can go, you simply run the risk of invalidating evidence if something shows up in a picture. If you need to smoke, then coordinate smoke breaks so that everyone with you takes a break. The investigation should stop during the break so as to eliminate the risk of capturing smoke in a photo. Also be cautious of other sources like your breath, etc, that may end up in a photo.

A dark vortex? Photo from a local cemetery near Port Jervis, New York. Digital image.

The photograph above is a digital image I took showing what appears to be a dark vortex in front of an old grave. It was taken in daylight hours. This vortex even has a name: Sony. Sony, you see, is the name printed on the camera strap on my digital camera. I created this photo and several others on the following pages to illustrate how simple explanations can have paranormal appearances. This picture was intentional – I was trying to photograph my camera strap. I wanted to show that camera straps can often be easily mistaken for paranormal anomalies.

Many seasoned investigators will tell you to remove the straps, because no matter what you do, it will get in the way at some point. If you take dozens of photos at a site and later find one with this type of anomaly, you may easily mistake it for paranormal activity.

We leave our straps in place and keep them clear of the lens by not allowing the camera strap to ever hang freely on the camera. If it is not around the neck, then the strap is wrapped around the wrist and hand. You can also look closely at this vortex picture and see that it looks like a form of material. Some camera straps are thinner and may not show as much detail. But chances are, if you come back with a digital image of a vortex, look at your camera strap. Chances are good that you will have found the source of your vortex.

A glowing vortex in the foreground and a transparent vortex off to the left of the grave stone. A paranormal jackpot? Digital image taken at a local cemetery near Port Jervis, New York.

The image above is a white strap that got a little too close to my camera lens while taking this nighttime photo in a local cemetery. The flash causes the strap to take on a mysterious, unearthly glow. To the left of the grave stone is another more transparent vortex. Both are caused by the same phenomenon: the strap getting in the way of the flash and lens. This photo was intentional and the glow is a clue as to the earthly origins of the vortex. However, had the strap been just out of lens-view it may have still reflected light and caused the transparent vortex as seen in the photo.

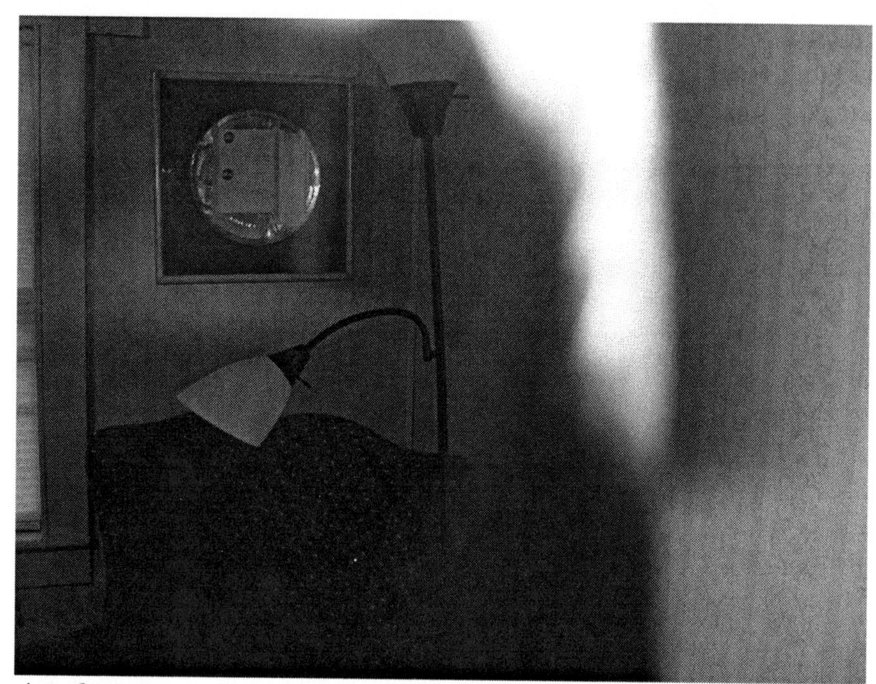

Another vortex like image captured at a residence near Port Jervis, New York.

Another example above shows how a vortex can show up in a digital mage taken indoors when the camera strap decides to make a cameo appearance. In this particular photo, the strap has taken on luminosity as well as casting a shadow into the picture.

A photo such as this, if it were one amongst a hundred, may cause you to mistakenly believe that there was something going on and you captured it! For most investigators the truth is not as exciting as the paranormal explanations, but the truth is the only acceptable explanation that we can accept in this field!

APPARITIONS

Digital image taken at the very abandoned, and very haunted, Tamarack Lodge in Greenfield Park, New York.

Quick quiz: What does this photograph show?

A. An unearthly visitor clinging to the fond memories of a crumbling resort.

B. A tortured soul wandering the grounds seeking revenge from the other side, raising a phantom knife in a homicidal rage.

C. My own shadow getting in the way of a daytime picture.

If you guessed A or B, you fail! OK, just kidding. Both A and B would be wonderful explanations if they were true, but letter C is the obvious correct choice

Shadows are easily created and this picture was taken outdoors on a bright, clear day. You can see from the shape of the shadow I am holding a camera up to take a picture. Note that shadows can be thrown in all manner of lighting conditions, and even in dark conditions. Flash cameras will cause shadows when the flash goes off. Infrared cameras will also cause shadows by their infrared illuminators. These shadows will not be visible to the unaided eye. However, they may easily be picked up by others with infrared cameras. This is something to keep in mind when taking images at haunted locations. You should always know where members of your investigative team are, and know where sources of light (including infrared) are coming from. It will help you correctly interpret your photos at a later date.

Digital image of a shadow figure at a local cemetery near Port Jervis, New York.

The cemetery picture above is another example of my shadow popping up in an otherwise good picture. It was taken with the sun to my back, casting a shadow onto the ground. The arm that is raised up and to the side I did on purpose, to add contrast to the photograph. Taking the time to analyze your pictures will help you weed out those that are simply not ghostly in nature.

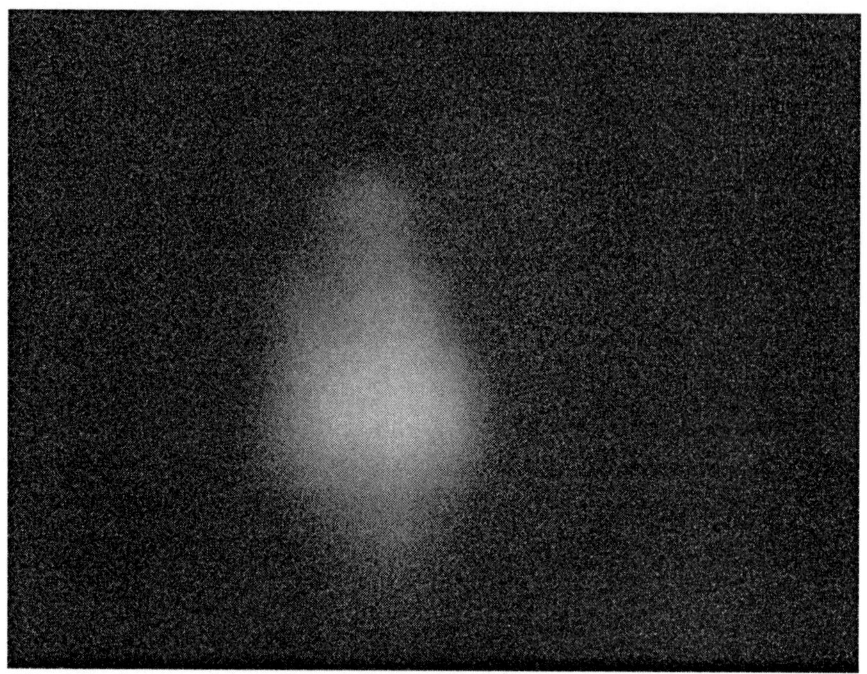

Ghostly image from the Tamarack Lodge, Greenfield Park, New York. Digital image taken in infrared mode.

The photo above was taken outside at the Tamarack Lodge. I did not notice this anomaly at the time that I took the photo. It was a day or two later that I discovered it while downloading pictures from the camera.

On its' face this is a pretty convincing photograph. It looks like the shape of a human floating in the darkness not very far in front of me. This could be, to some, the Holy Grail of ghost hunting! I was certainly excited when this picture came onto the screen. Maybe, I thought, I had captured a ghost on camera!

But to stop there and declare the photograph to be a genuine image of a ghost would be unscrupulous and unprofessional. So I made a note of the photo's number and continued on to the next. I did not have to analyze too much to find the very simple explanation.

Linda Zimmermann, the Ghost Investigator, with an EMF meter outside of one of the abandoned buildings at the Tamarack Lodge.

My ghost was none other than Linda. The picture above was taken just seconds after the image with the human-like apparition. You can see when examining both photos that the shape of the apparition and the shape of Linda with her vest on are nearly identical.

The photograph of the apparition was the result of the camera's auto focus having difficulty finding a target to focus on. When the shutter opened it was at a moment where the camera had simply not found an adequate focus. Instead of Linda with a meter in her hand, I now had a ghostly apparition floating in the darkness.

This is why looking at photos with a critical eye is so important. Sometimes the photo itself may not give up its secret until you look at the other images taken at the same time.

To better illustrate how you can make the comparison I have cropped the apparition and Linda on the next two pages.

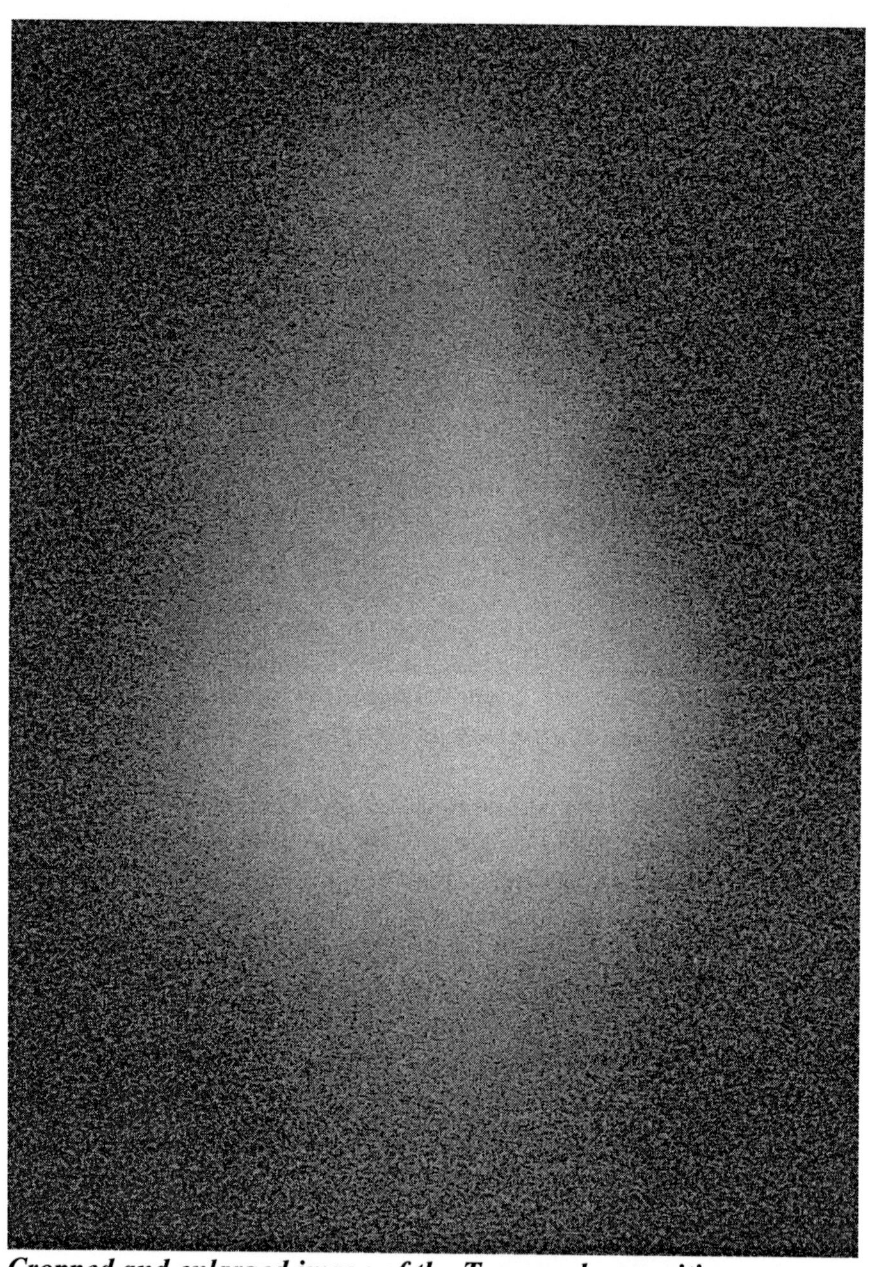

Cropped and enlarged image of the Tamarack apparition.

Take a good look at the apparition, especially the shape.

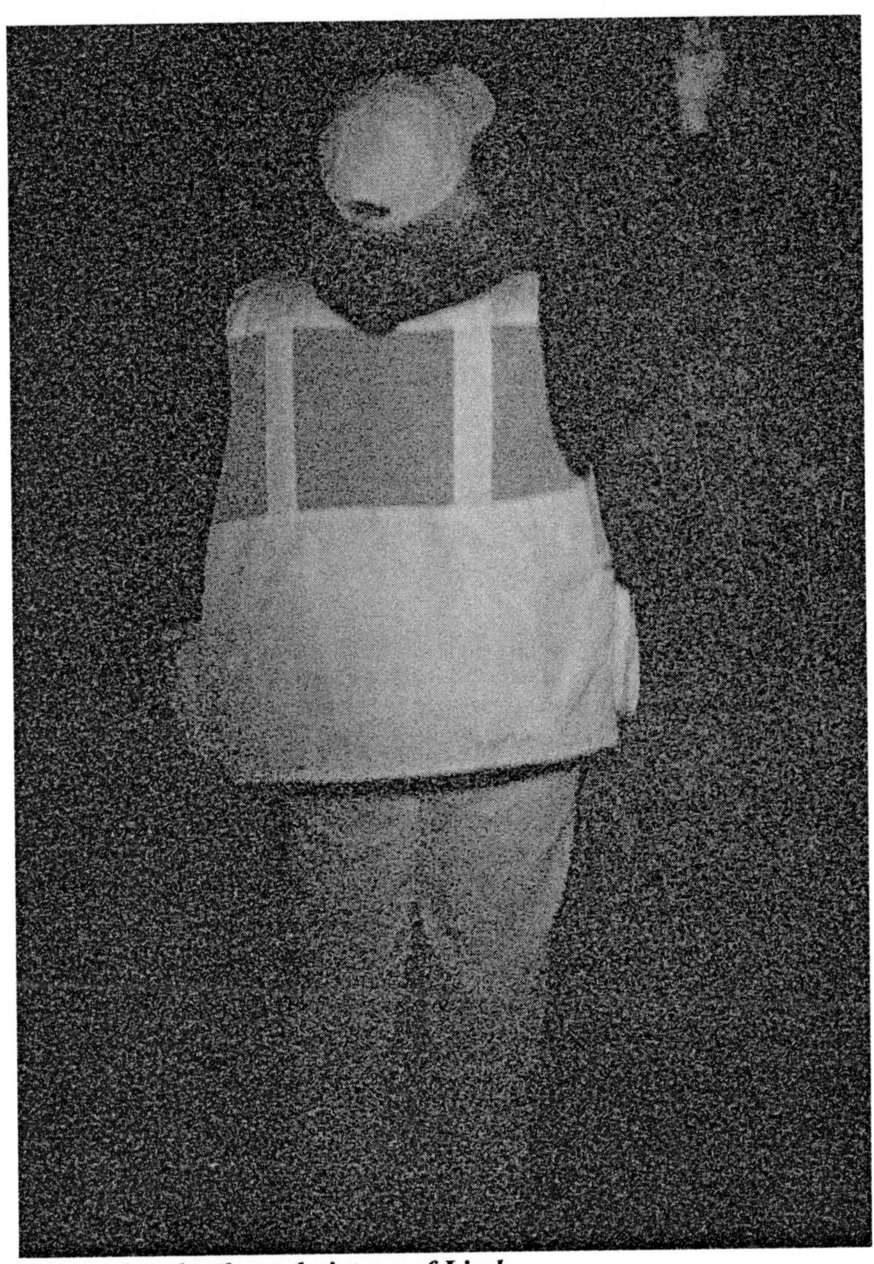

Cropped and enlarged picture of Linda.

The comparison is easy to make. No ghost, but honesty and integrity are critical elements of ghost hunting. Be honest with your photos and let the truth speak for itself.

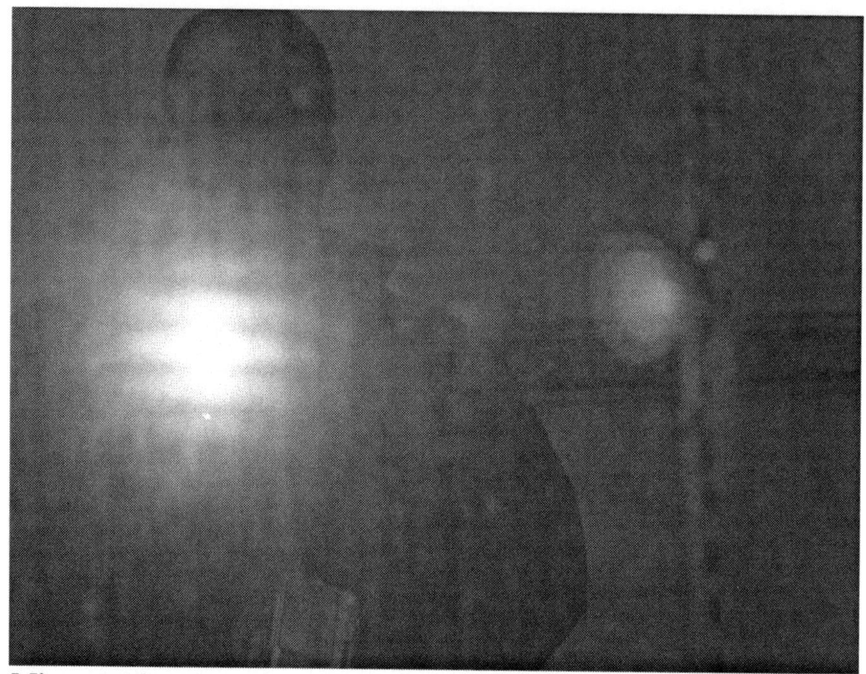

Mirror, mirror on the wall, who is the best ghost hunter of them all? My reflection in room D-10 at the Tamarack Lodge.

The picture above is an example of what mirrors and flash create when they combine. We were investigating room D-10, which is said to have an apparition that can be seen in the mirror. I should have known better than to take a flash photo into a mirror, but I went ahead and took it anyway. Blame it on the mold in the Tamarack!

It is pretty obvious that there is a flash source in the photo along with me (that would be the good-looking guy in the mirror, hahaha) and also some odd white reflection off to the right. Mirrors and windows, and most any other source of glass in a location, will cause you problems. They will reflect light from flashlights, camera flashes, passing cars, street lamps, and any other number of sources that are all very normal. You will capture odd shapes and odd reflections due to these tricks of light, many of which you may not notice at the time the picture was taken. It is hard to accept anomalies under these conditions as being paranormal, so check, double check, and triple check them before you post them as the real deal!

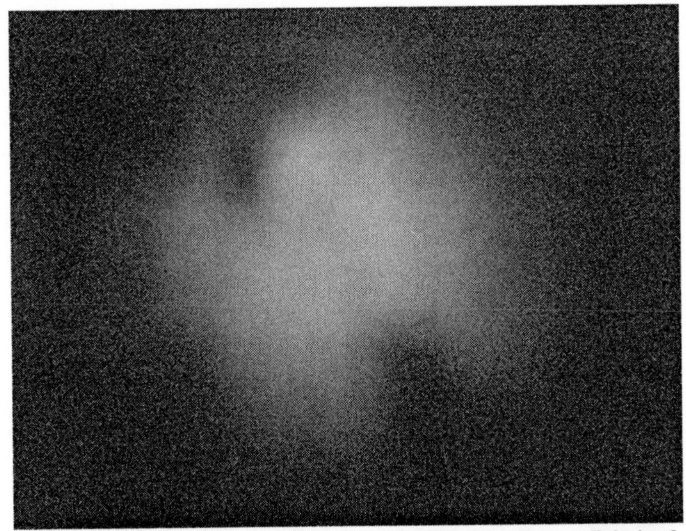

Apparition? The Columns, Milford, PA. Digital image in infrared mode.

Photo taken seconds later shows the apparition to be a mannequin. The first image is out of focus.

The photos above are shown to drive home the point that you have to carefully look at your photos. Do not assume because it looks like a ghost, it is a ghost. While finding the earthly explanation can be disappointing, your credibility is preserved and you have a reason to go out and keep on ghost hunting.

Know where others in your group are. The picture below was taken in an old cell block at the former Ulster County Jail in Kingston, New York.

One of our guides at the old jail was standing half in the doorway to this cell block when I snapped the picture.

Infrared digital image taken at a cell block in the former Ulster County Jail. There is an outline of what appears to be a head and torso in the area where the door is.

I included the photo above because I don't know for sure what is actually depicted in it. There appears to be the outline of a head and torso in front of the door. I can not tell you that this is a picture of a ghost. It could be a shadow cast by one of our infrared illuminators on a camcorder. It could be an illusion created by the pixels in the image.

Pareidolia is a phenomenon where we see familiar and recognizable objects in things. Usually we see faces (like orb photos with a smiley face in them). My photo above could just be an example of pareidolia. I will leave it up to you to decide.

Keep in mind that as you look through evidence, be cautious not to see things in the picture that are simply not there. Pareidolia is unavoidable – it seems to be hardwired into our brains, but that does not mean that we can not critically evaluate a photo and call it for what it is.

The sequence of images below was taken at a cemetery near Port Jervis, New York, using a digital camera in infrared mode.

Photo 1 shows the grave with no apparent paranormal activity. Photo 2 shows the same grave seconds later with an eerie mass in the lower left corner of the image. Photo 3 shows the same grave another four or five seconds later with the mass becoming quite large along the left side of the image.

This sequence of photos, however spooky, is far from paranormal. While adjusting the camera zoom and focus on the camera I managed to capture the tip of my finger in the second photo and a bit more in the third. My finger takes on the eerie appearance due to the close proximity of my finger to the lens and the infrared illuminators on the camera.

When taking photos always be conscious of where your fingers and hands are in relation to the lens. They can easily get in the way and later turn up as eerie-looking apparitions.

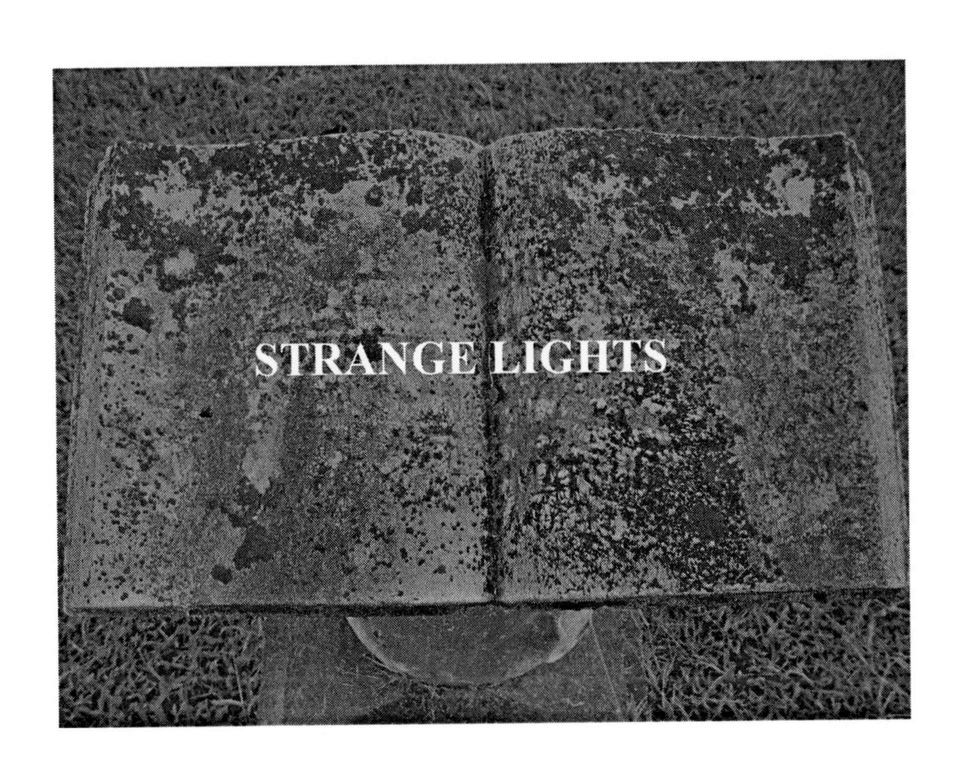
STRANGE LIGHTS

Years ago Linda and I had the opportunity to accompany a group of young ghost hunters to a cemetery in Rockland County, New York. After the sun had set we were noticing that in our flash photography there was a strange glowing red ball of light in the distance. Now we always try and maintain our professional composure, but there is something very unsettling about seeing something that you believe may be paranormal while in a dark cemetery. Yes, the red light was giving us the creeps, but we took the time to investigate further and determine that the red light was the reflection of our flashes off of a red reflector attached to a tree.

This leads me to discuss cemeteries as problematic locations for investigations. Yes, they are generally considered by most paranormal researchers to be a good place to find ghosts. Some spirits for some unknown reason attach themselves to cemeteries. Perhaps it is the connection to their earthly body that keeps them there. Perhaps it is the energy of loved ones that visit graves, or the energy of mourning that comes with death, burial, and last goodbyes.

In any case, cemeteries are good places to look for ghosts. But they are bad places to obtain good photographic evidence. You have seen in the preceding pages how easy it is to misinterpret cemetery photos, and I have not even included pictures with phantom-looking balls of light that are nothing more than moths, mosquitoes and other flying insects. Let me tell you that an insect observed close to the lens in infrared mode looks like something from the other side. They glow, move in odd manners, and because they are so close to the lens the focus is off and you cannot easily make out that it is a bug.

This is not a problem limited to cemeteries--any outdoor location is prone to bugs, dust, etc. Several years ago Linda, Bob, I and a friend of mine visited Gettysburg, PA. It was my first time there and it was a moving experience. Of course, Gettysburg is reputed to be very haunted and we went out the first night to check out some of the battle sites. One location in particular that night had a circus-like atmosphere. It was dark and people were sitting on lawn chairs looking out into the field. Someone was playing music and people were holding their cameras up and taking flash photographs. After each flash you could hear someone exclaim that they had "another one," meaning that they had another orb on camera. Now we have already covered orbs that are the result of dust, moisture, etc., but what these ghost hunting enthusiasts were capturing was a different type of orb: fireflies! The field was loaded with fireflies. They were everywhere and when they would light up the field would look like strings of Christmas lights were strewn about. So it was not a shock that they were capturing orbs. But it was a shock that they didn't make the obvious observation that the fireflies were the orbs that they were catching.

Digital image taken in a cemetery outside of Port Jervis, New York. The thin streak in the left center of the photo is some type of bug. You can also see reflections off of polished surfaces of nearby grave stones. The odd glowing mist in the center is the result of the flash reflecting off moisture as the cemetery is bordered by rivers on both sides and there was a light fog hanging in the air.

Strange balls of light in a cemetery photo taken with my digital camera.

The photo above shows how lights from surrounding street lamps and homes can show up as supernatural balls of light. The cemetery that I took this photo in is quite large and it is easy to forget which section a particular photo was taken in.

If you are at an unfamiliar location and take a lot of pictures, you may end up photographing something like this. Often days will pass before we get to download pictures from the camera to the computer for evaluation. If you are not sure where you were, then you may just end up calling this type of picture paranormal when it is simply normal.

If you do outdoor investigations, get there in the daylight to allow plenty of time for a thorough examination of potential sources of reflection. Also look for the location of adjacent sources of light that may show up later in nighttime images.

Demonic eyes staring out from the dark abyss, waiting to devour human souls?

The picture above shows how memorial candles placed near a grave can take on a demonic glare under the cover of darkness. This image was taken with my digital camera in infrared mode. The focus is on the large cross to the left. This results in the candles, which are more distant, being out of focus. The combined result is a spooky photo, but not something that is paranormal.

The pictures on the following pages highlight two other problems with cemetery investigations: animals and the living. Animals, such as this ground hog, not only create large holes that can trip you, but they may end up in your photos as eerie specters rather than furry animals. And the living that sometimes hangs out in the cemetery at night may be far more frightening and dangerous than any of the restless souls who may wander about.

I am a police officer, so I generally have some type of protection with me. The bottom line is do not go to these places alone.

A furry critter popping his head out of a hole in the ground at a cemetery near Port Jervis, New York.

This empty beer bottle is a stark reminder of what goes on here at night. The living are generally more frightening than the dead!

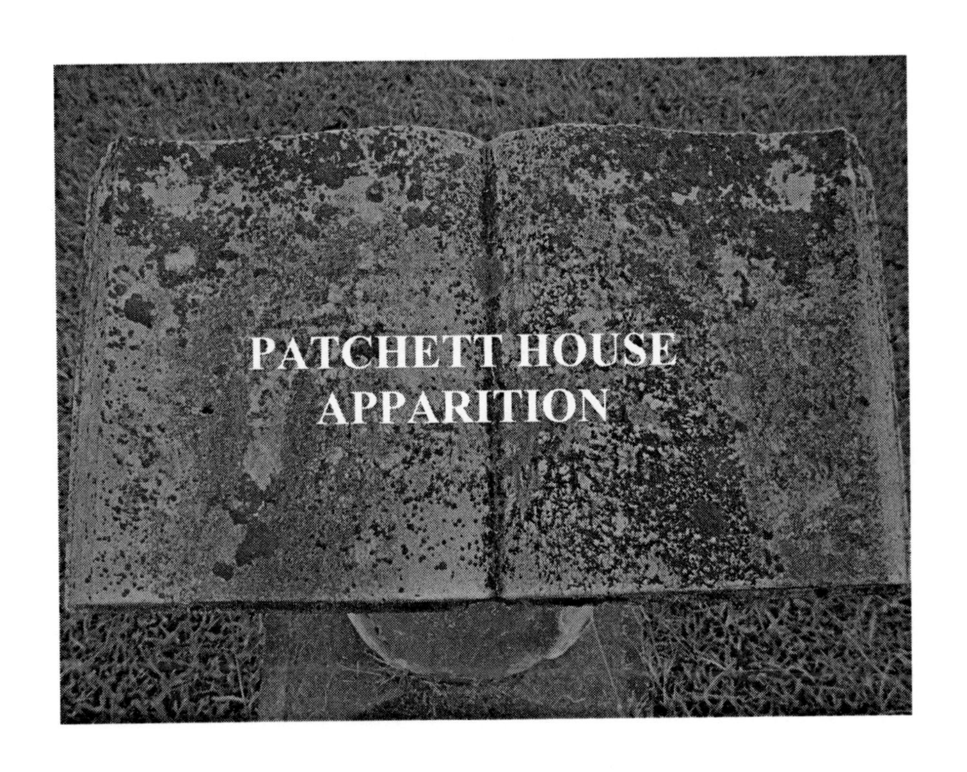
PATCHETT HOUSE
APPARITION

The Patchett House was the first investigation in which we used a trail camera. The camera we use is a Stealth Cam with built in infrared motion sensor. The camera is designed to take color pictures in day light, and black and white infrared pictures in darkness. Hunters use these cameras to capture images of game along trails and in the woods to determine prime hunting locations.

The Stealth Cam was set to take nine consecutive digital images when the infrared motion sensor detected movement. The camera would take nine pictures in a sequence then rest for minute before rearming the motion sensor. The pictures taken in sequence would be a few seconds apart.

The images that we documented from the Patchett House are startling. The camera was set up on a tripod in the former embalming room located in the basement. There were three of us in the building during our return investigation: myself, Linda and Barbara Bleitzhofer. During the investigation we were on the second floor when we hear a very loud crashing sound from the floor below us. It was so loud and distinct that my first instincts were someone had entered the building and I worried more about a burglar or other intruder than I did a ghost. After a thorough search of the ground floor turned up nothing, I suggested we check the basement to make sure that the trail camera did not fall over. We turned on the lights – at my insistence – and ventured into the basement. Along the way through the basement to the embalming room I had set up several motion alarms. I have gotten into the habit of using them for two reasons. One is that they can alert you to paranormal presences and we have had them go off unexpectedly on investigations. The second is to alert to the presence of other people who may come into a location. This way there are no surprises during the investigation and you do not capture a photograph of a living person who may have inadvertently walked in on your ghost hunt.

The camera was quite intact and still in location and we resumed our investigation. A few days later I downloaded the images form the camera and started to go through them. The first nine photos captured were amazing and showed something was in the room with the camera – and had *touched* it!

The entire series of nine pictures is printed on the following pages.

Image #1. Nothing is visible in the photo that would account for having triggered the infrared motion sensor.

Image #2.

Image #3.

Image #4.

Image #5.

Stealth Cam 04/15/2009 21:14:15 ● 55F

Image #6.

Image #7.

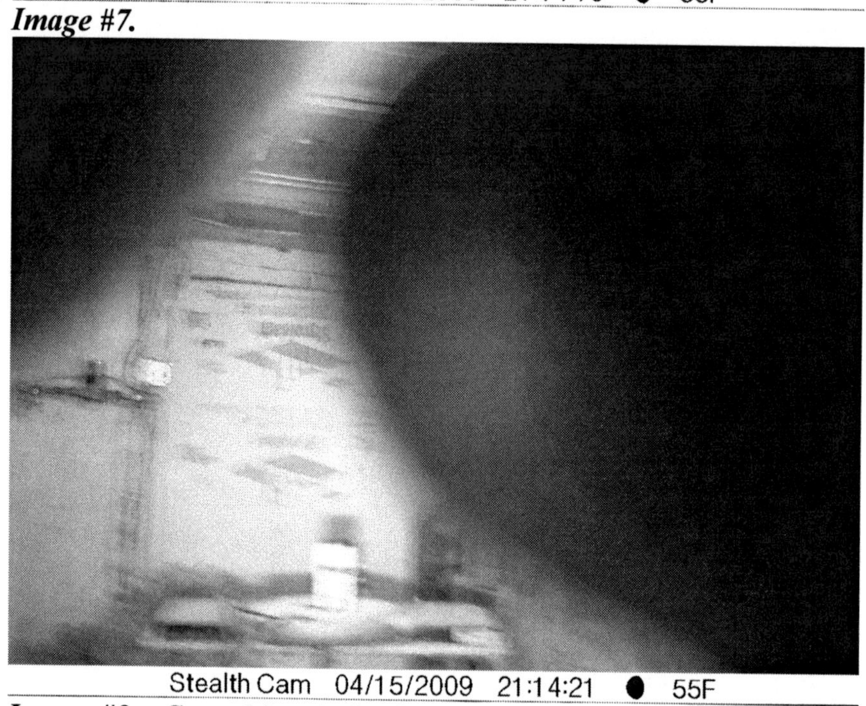

Image #8. Something is in front of the camera and the image appears to blur as if the camera is starting to move.

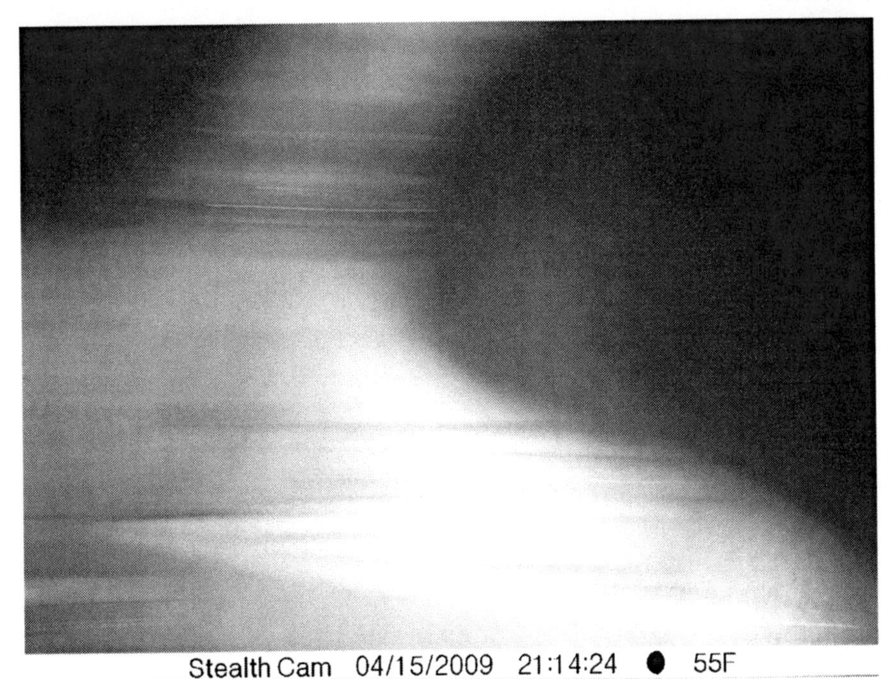

Image #9. The image is completely blurred with movement.

Something triggered the trail camera infrared sensor and the camera began to take a series of nine photos. Thirty seconds later something is grabbing the camera and it appears that the camera is moving. It is in black and white because the images were taken in infrared mode. We were all accounted for when the pictures were taken and did not trigger the camera while in the dark.

What moved the camera in the basement of the Patchett House? I attempted numerous times to recreate this photograph using my own hand. I could get some close results; however, my fingers looked lighter. This is expected as the fingers would catch more of the infrared illumination when the camera was taking a picture. I tried different angles but could not replicate the same anomaly as in the sequence.

The next page shows how objects too close to the trail camera appear. The first is how I look when I get a little too close the trail camera in the basement of Iron Island. The second is a photo of my fingers when I tried to replicate the Patchett House anomaly.

Infrared image. The closer I get to the camera the brighter I get.

My fingers grabbing the trail camera. The time of the camera is off – this picture was taken in the dark in infrared mode.

POMONA HOUSE
APPARITION

Linda captured this astonishing photo at a private residence in Pomona, New York. Photo courtesy Linda Zimmermann.

The photo above is shocking. Linda took this photo on an investigation at a residence in Pomona, New York, after seeing a something pass by this area in the room. It is detailed in *Ghost Investigator Volume 6; Dark Shadows.* A lot of time was spent analyzing the picture including determining who was there and where they all were at the time the photo was taken. Nothing outside of the residence could account for the apparition.

Linda revisited the location and was able to determine that the figure in the image stood about 5'8" in height. The photo on the following page shows the same room in daylight. It is clear to see that there is nothing here that could be mistaken for the shape of a human. Whatever Linda had captured in this photo was not of this world. To date it ranks up there as one of the best pieces of photographic evidence of the existence of ghosts.

The living room in day light. The figure would be standing to the left of the small table with the sculpture on it. Photo courtesy Linda Zimmermann.

When you take the time to be critical and evaluate your photographic images, you enhance your credibility. A picture like the one Linda captured of the dark apparition then becomes much more valuable as evidence of the paranormal. Take your time and review your pictures and make logical, well thought out conclusions. You will become sharper and may be rewarded when you capture an age that defies all explanation.

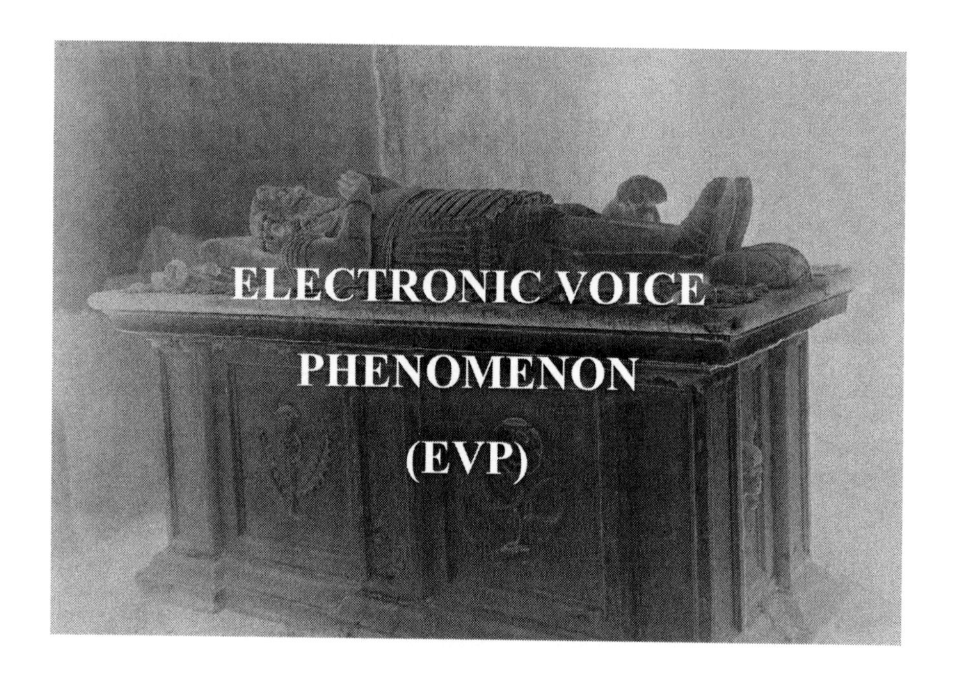

"Nobody knows whether our personalities pass on to another existence or sphere but it is possible to construct an apparatus which will be so delicate that if there are personalities in another existence or sphere who wish to get in touch with us in this existence or sphere, this apparatus will at least give them a better opportunity to express themselves than the tilting tables and raps and Ouija boards and mediums and the other crude methods now purported to be the only means of communication."

- Thomas Edison

Electronic voice phenomenon, or EVP, is a popular form of evidence sought by ghost hunters. If a picture is worth a thousand words, then an EVP is worth as many or more. For many claim that the sounds recorded in an EVP are that of ghosts and spirits! I like to go a step further and use the term electronic audio phenomenon, since not only voices are picked up. However EVP is the accepted term in the field so we'll stick with that. EVPs are typically recorded but not heard at the time that they were recorded. This is not always the case and Linda and I have both captured EVPs where we heard it and recorded it at the same time (think about the dying breath from the Van Winkle House from the chapter about residual haunting).

EVP work is quickly becoming a central part of many ghost investigations. EVP sessions, as many refer them, are directed attempts at asking questions and then pausing to allow for a response. A response from the other side, that is. During these sessions you typically do not hear the response with the unaided ear. However, upon playback of the recording medium, you can hear voices and sounds. And when these voices answer in direct response to a question you ask it is compelling evidence.

Most investigators today use digital recorders for EVP work. Like digital cameras they are a logical and economical choice. You can record a significant amount of high quality audio on a small medium and even download it to a computer for easy review and storage. There are no tapes to fuss with and no mechanical parts to cause background noise. We use them for both EVP work as well as set up at strategic sites within a haunted location. I am sure that someone out there is still using a good old tape recorder for EVP work. Some have theorized that EVP imprints itself directly onto the magnetic medium of the cassette tape. Interesting theory, but yet digital recorders still capture EVP with no magnetic medium. I don't think cassettes put you at an advantage as much as a disadvantage. Cassettes need to be new every time to avoid any type of artifact recording. The gears produce background noise. And most cassette recorders are bulky. They may be practical for recording interviews and making initial notes aloud on a place. For EVP work I think they are best relegated to the bottom of the bag.

In the chapter on personal experiences I wrote about my encounter with the ghost of a small boy at the Iron Island Museum in Buffalo, New York. During this encounter, in direct response to my question asking the boy's name, a voice was recorded saying 'no'. It was a quiet, child's voice. It was not the voice of anyone in the room at the time. There were no children even present. What could account then for the voice? I believe the ghost of that boy. Given the totality of that encounter - sudden electromagnetic field fluctuation, the cold touch on my hand and the recorded voice - I believe we had an encounter with a genuine spirit.

Unfortunately, not all EVP evidence is reliable, or even credible. On a positive note, I do not believe that most people purposely fabricate EVP evidence. I don't believe that most people fake any evidence for that matter. It happens and when it does legitimate ghost hunters suffer. But by and large most ghost hunters are honest people looking for good evidence of the paranormal. So how do good ghost hunters produce bad evidence, including EVP? Misinterpretation is the culprit.

Most people have seen one of the popular crime shows on television where a crew comes in with sophisticated equipment, scans a crime scene and gets all manner of evidence, and then goes to the lab where even more sophisticated equipment identifies who the evidence belongs to. Fingerprints and DNA are located, collected and identified in almost mythical style, and all within one hour. Shows such as this have made the work of real police officers more difficult as people expect us to have these marvels of science available to us at our disposal. I wish we did. I wish I had the ingenuity to invent one so that I could retire a zillionaire.

The reality is that evidence at a crime scene is not often so easy to locate and identify. Do you have any idea how many fingerprints are in your house or apartment or workplace? Fingerprints are everywhere. Do you know how many skin cells you shed every day? How about how many skin cells that other people shed every day? DNA is everywhere. It is up to a criminal investigator or crime scene technician to know where to collect evidence at a crime scene and how to preserve that evidence so a qualified expert can properly evaluate it. That is how conclusions are made (e.g. the fingerprint on the murder weapon belongs to the defendant).

In ghost hunting we are both technician and evaluator of evidence. We collect the evidence at the scene of the haunting and then later evaluate it for evidentiary value. In the case of EVPs, this means listening to recordings for anything that is unusual. Once you have identified that unusual recording you need to evaluate it. Collecting it is the easy part. Evaluating it is where the problems come in.

If you have never heard a purported EVP, go to a popular search engine of your choice and type in EVP. Plenty of options will be available to you. Find a site and listen to some of the EVPs that are presented. Generally you will find EVPs listed with quotes as to what the EVP purports to say. Listen to them, and listen more than once. Chances are you will hear things that will make your hair stand up. Now after you have a few EVPs under your belt try this: go to another site and find some EVPs but *do not* read what the EVP is supposed to be saying. Try and determine what it is that you are hearing. Listen to it several times and write down your conclusions. Then see what the website claims the EVP is saying. Did you draw the same conclusion? If not, listen to it with their suggestion in mind and see if you now hear it. Do this with a friend and take turns, it is fun and educational. Far too often EVP evidence is sketchy at best and you only hear something after reading or being told what you are supposed to hear.

If I tell you I have an EVP of a man saying "I died here by murder" and then I play that EVP you are probably going to hear a man say "I died here by murder". In reality that EVP may be nothing more than a static with background noise that is otherwise unintelligible. The power of suggestion is powerful and not to be taken lightly when it comes to EVP.

And EVPs are not problematic simply by the power of suggestion. Noise is everywhere. One of the reasons investigating at night is a good idea is because it tends to be quieter. I do not think that ghosts are necessarily more active at night, just that they are more easily heard or seen. You are less likely to be distracted by extraneous noise during evening hours and are therefore more likely to be aware of things that are occurring. What kinds of noises can be misinterpreted as EVP? Almost anything, from breathing, to movement of clothes, passing vehicles, insects, talking and normal sounds that occur in a home (the furnace kicking on for example).

If you take your EVP work outdoors to a cemetery, battleground or other location you compound the problems. Sound travels and can travel quite far. It becomes very difficult to ascertain whether an EVP is truly something from the spirit world, or just human activity picked up by the microphone on your recorder. Even indoors sound can travel through walls and be picked up only to be misinterpreted later on.

How do you solve this problem? I am not sure I can offer a simple solution. One of the best things you can do is try to minimize the noise that you create during an investigation, especially during an EVP session. Coughs, clearing of the throat, sniffling, movements, banging into to objects should be made note of. It is easy enough to do. "I coughed" said aloud after coughing will clearly identify the cough as being non-paranormal in nature. Like photographic or video evidence, you may not get around to listening to your recordings right away. For most of us, ghost hunting is not a full time job and we do not have the luxury of unlimited time and resources necessary to review potential evidence immediately after we obtain it. Days, maybe even weeks, may pass before you sit down and begin to listen to your EVP sessions and other recordings. You're unlikely to recall every noise you made so saying them aloud at the time you or someone else made them will provide for fresh and immediate recollection.

Another easy step is to control the EVP sessions by documenting everyone present, where everyone is seated, and sources of potential sounds. If everyone says their name aloud at the beginning of the session you will know who is there and also know from which direction their voices are coming.

The simple procedure is to then ask a question and pause, allowing a short period of silence after the question before asking the next question. When you listen to the EVP session you are going to focus on listening for answers to your questions. That does not mean ignore other potential sounds. But if you ask aloud a question such as 'what is your name?' and hear recorded 'Bill", then that is going to make for a better EVP than hearing what sounds like the name 'Bill' mixed into a conversation that you and another investigator are having.

When evaluating EVP evidence listen closely and when you find something that you suspect is paranormal in nature review it, over and over again if necessary. Ask other investigators to listen to it *without* telling them what you think it is. Try and determine if the origin has a logical explanation and if it does then discard the EVP as evidence. Practice makes perfect and the more you do this the better you will get at it.

My experience also shows that the best EVPs are those that are obtained without the use of audio enhancement software. We like to be able to hear the EVP with our own ears – not after the sound has been processed. For me, the best EVPs jump right out at you and leave no question as to what they are saying.

We also stay away from white noise generators. When looking for EVPs the last thing you want to do is generate more noise! While many have claimed that these devices work well at obtaining EVPs I think it just raises doubt as to the veracity of anything recorded being supernatural. There are also devices being used now under various names that use white noise generators and AM receivers to capture purported ghosts. I am not here to offend anyone, but the moment I start to hear AM (think AM radio) receiver I start to think of interference. If you scan the dial of your car radio fast enough you will pick up snippets of audio. These devices also require processing of the recorded audio in order to really hear anything. For me these devices are just to fringe and open to potential sources of interference. It doesn't mean that extensive research and documentation won't show them to be reliable and accurate means of communication someday. But for now, I'll stick with what works and what I can reasonably say has produced some good EVP results.

Another area of controversy in paranormal research is in provoking spirits. This is where the investigators verbally harass or attempt to stir up the spirits in an attempt to get the spirits to respond. To some, provocation is disrespectful and a practice that they avoid. Others find it to be a useful tool. I used it with some success at the Iron Island Museum in Buffalo, New York. After my moving encounter with the spirit of a child I had felt that a negative, overbearing male energy had driven the child from the room. I don't know if it was my paternal or police instincts (or perhaps both) that kicked in, but I wanted to go after this male presence. And talking nicely to him was not the way to approach him. In this case being firm, commanding and borderline insulting ('hey you can pick on a little kid but not grown adults') actually managed to drive the spirit from the room and back into the second floor attic area where it is though that he resides.

Provocation is not a black and white issue in ghost hunting. There is a vast gray area – the gray area of personality. Provocation may be appropriate in some instances while being totally inappropriate in others. Each investigation is different as are the entities involved. You have to evaluate the haunting and the personalities of the spirit involved to see if provocation is appropriate. If the spirit is a negative presence such as I felt at Iron Island then provocation may be appropriate. If you are encountering the spirit of a meek, mild mannered person then that spirit may simply avoid you all together if you begin provoking him. The spirit of a child may not respond well to provocation and may require a gentler, parental approach to communication. The bottom line is that you have to allow the unique facts and circumstances of the investigation to guide how you operate and how you attempt to communicate with spirits. There is no one size fits all approach to ghost hunting and certainly when it comes to communicating with them!

Keep in mind also that you may be addressing a spirit for whom it is just wrong for you to even be addressing. While we were investigating the Cliff Park Inn in Milford, PA, manager Stephanie Brown told us of a story about room #10, which is called Sally's room for the resident spirit of Sally that is said to reside there. She recalled that a group of paranormal investigators where in this room conducting a ghost hunt and one of the male members of the group was laying on the bed attempting to get Sally to give him some type of sign of her presence. He was not getting any results. Stephanie pointed out to him that he as a man – a strange man – in a woman's bed, and asking her to make contact, and he shouldn't be surprised that she wasn't coming around! So think about whom you are addressing and determine if they will even address you back.

Another issue to address is language – and not the vulgar type. I have watched television programs featuring ghost investigators at foreign locations attempting to gather EVPs. The problem is that they were in countries where English is not the native language and they are trying to communicate in English! If you are in a Romanian castle trying to contact the spirit of a long dead Romanian Prince, do you think that English is going to be the language of choice? Perhaps you ought to have someone speaking Romanian. Now I am the most monolingual person that I know. I have been slowly teaching myself Finnish for several years now and honestly can speak, read and understand a little. Not enough to survive on yet, but I am still learning! So I have a healthy respect for people that can speak multiple languages. But I do not think that somehow you cross over to the other side and become fluent in all of the world languages. It is my opinion that the language or languages you spoke in life will be those that you understand and communicate with in death. I was talking with Linda recently and she was telling me about a haunted location where there were English, French and Native American soldiers alleged to haunt the place. In a place such as this you may expect to encounter a spirit that does not speak English, so your EVP sessions may have to take that into account. You as the investigator need to adapt to changing circumstances of an investigation. Don't expect the spirits to do that for you.

"For every problem, there is one solution which is simple, neat and wrong. "

- Henry Louis Mencken

Imagine this scenario: you have just returned home from a week-long vacation to find that your home has been burglarized. Your jewelry, television, and gaming system are all gone. You call the police and wait for an officer to come and take your report. When the officer arrives to your home he walks up to you and announces with bravado "I am here to prove that your house was not burglarized." Get on the phone with the dispatcher and ask for another cop!

If the above scenario sounds ridiculous, it's because it is. And it is just as ridiculous for a paranormal researcher to show up at an alleged haunting and proclaim that they are there to prove that the place *is not* haunted. To be fair, it is also just as unprofessional to show up with the intent to prove that a place *is* haunted.

Both extremes introduce an element of bias to the investigation. This bias can lead to "confirmation bias", which is where we tend to find and interpret information in a manner that is consistent with our own beliefs. If you are at a haunted location with the belief that you are going to disprove the haunting, then you are going to potentially introduce an element of confirmation bias into your work. This is no different than overzealously wanting to prove a haunting and introducing that same bias. In either case the potential is that evidence may be misinterpreted to conform to either point of view.

Now debunking is not necessarily always as extreme as trying to disprove a haunting. Debunking is a practice in paranormal research wherein investigators try to disprove different paranormal claims and experiences. Many groups have debunkers amongst their ranks. I have been asked on several occasions during presentations if I debunk during my investigations.

What is my problem with debunking? Well, debunkers may run the risk of introducing bias into the investigation by seeking to disprove alleged paranormal activity. For example, if it is reported that a door opens and closes on its own and the debunker can easily replicate this, then the activity will have been declared debunked. Now just because you can replicate it doesn't mean that it also is not influenced by paranormal activity. For example, just because the floor boards creak when I walk on them does not mean that you have to rule out all creaking floor boards on an investigation as being non-paranormal.

But debunking is more than a skill that some researchers employ. It is almost an investigative method. Essentially, if something can be debunked it is ruled out as paranormal. That which can not be debunked is then considered as potentially paranormal in nature. If this was a medical text and we were talking about medical diagnoses we may use the term *trashcan diagnoses.* Since it is a ghost book I will co-opt this word and call it *trashcan investigating.* You go in, debunk all that you can and whatever maybe left over you then evaluate it as potentially paranormal in nature. If this is how you investigate and it works for you, then keep on with it. It isn't for me.

How do we investigate? We go in with an open mind and objectively seek the truth. We ask questions, we gather reports, we look into what people are experiencing, we explain what can be explained, and we gather evidence – both objective and subjective evidence. We do not discard personal experiences as valid evidence. We take the totality of the investigation and draw our conclusions from there. Every investigation is slightly different and the outcomes are always unique. You have to be dynamic and go with the investigation and how it is unfolding. We never employ a rigid protocol because hauntings are unique and vary in complexity and experiences. We have gone on investigations where we use nearly every piece of equipment, while on others where we barely use any. Every case is different. We adjust accordingly and have attained good results from that.

If you are going to a haunting with an overzealous need to prove the existence of ghosts then you are doing a disservice to all of us. Yes, you can and should have a basic belief in ghosts. It seems silly to be a paranormal researcher who doesn't believe in them. But if you are so convinced that there is a ghost in every corner and an orb in every picture (even one with a happy smiley face) then you do us a disservice. I have to borrow a line from my good friend and amazing psychic, Lisa Ann: just because your light flickers doesn't mean you have a ghost – it may just mean you have to change the light bulb. Not everything you encounter on an investigation will be paranormal and you will find reasonable explanations for a lot of what you encounter. Finding these explanations is good investigative technique, not debunking.

On the other end of the spectrum, if you are so adamant that you are going to disprove a haunting then you may not see the proverbial forest through the trees. You may be so willing to discount any and all occurrences that you discount those that may have valid paranormal causes.

Good advice is to go into a haunting with an open mind and be receptive to what you experience. Evaluate your evidence with a critical eye and look for logical explanations. Do not discount sensory experiences and do not discount eyewitness accounts and the history of the location. Avoid confirmation bias and the tendency to go in and be the 'debunker in chief' and end up with a trashcan investigation. Good investigative skills and professional conduct will not only get you good results, but will enhance the credibility of the field.

"The most beautiful thing we can experience is the mysterious. It is the source of all true art and all science. He to whom this emotion is a stranger, who can no longer pause to wonder and stand rapt in awe, is as good as dead: his eyes are closed."

- Albert Einstein

We are fortunate to use a talented psychic named Lisa Ann on many of our investigations. Lisa Ann is a credible, competent psychic and her intuitive skills have added new dimensions to our investigations. If you are not currently working with a psychic as part of your team then you are missing out on a vital aspect of the investigation.

Many skeptics – and even long time ghost investigators – criticize the role of psychics in an investigation. There is often good reason for this. For just as there are unscrupulous ghost hunters there are unscrupulous psychics. Many simply believe that a psychic can do a little checking in advance of an investigation and go into haunted location armed with facts and information that will be startling and revealing in nature. That is a valid concern and one that we take seriously but never have to worry about.

But if you do decide that a psychic would compliment your investigations then follow some simple advice and precautions that we use to enhance both your credibility as well as that of the psychic.

To begin with do not tell the psychic in advance where you are going. We always meet Lisa Ann at a neural location ahead of time and car pool with her. We do not discuss the investigation at all. Not even a hint as to where we are going. There are plenty of other things about life for us to discuss and we manage to have some great conversations without ever giving her a clue as to where we are going.

The next most important step is to sequester the psychic from people related to the investigation but not part of your team. These are usually homeowners or their invited guests who want to see the psychic in action. By keeping them at a distance you eliminate the likelihood that one of them will say something that can contaminate the psychic's work.

A great example comes to mind from a family trip to Florida a few years back. At the Universal Studios Park was a psychic all dressed in Gypsy garb who talked us into having a session with her. Now it was Florida and hot – especially for me as I dislike hot temperatures unless it is in a hot tub or sauna! So I was wearing a tank top and shorts. During my session the psychic was able to determine my sons were twins, their exact age, birth day and birth sign. Now to the uninitiated this may be wonderful evidence of her powers. To me there was a more simple explanation: I had given her the information without even knowing it. First off she had seen me with the twins (they are identical) and then my right arm has their names and birth date prominently tattooed on it. Very easy for her to see it and come up with this revealing information. A bit of common sense goes along way and I accepted her reading for the entertainment value that it was.

But if you bring a psychic to a haunted location and the homeowner runs up to him or her and says “tell me that you see the dark figure in the hall” your investigation has been contaminated. It is better practice to keep the psychic separated and allow him or her to see that dark figure first, then to corroborate it later with the homeowner.

Make sure you don’t influence the psychic either. We have followed behind Lisa Ann wanting to jump up and down at the information she was validating. But that would jeopardize the value of her information. So we put on our best game face and leave the amazement for after her initial walk through.

Finding a credible psychic is probably your hardest task as an investigator. If you know one and like how he or she operates try integrating them into your investigation. Do not replace your own experiences and equipment with a psychic, however. The psychic should be viewed as yet another tool to help gather information and evidence of the paranormal.

"The quality of your evidence is inversely proportional to the number of investigators present at the investigation."

- Linda Zimmermann

I chose the quote by Linda to bring the book to a close and to cover some good tips for great ghost hunting (*in no particular order*):

I. The more investigators you have at a site the more likely it is that you are going to contaminate your evidence. The size of the location should be your guide. A place like Eastern State Penitentiary could easily accommodate a large group of investigators. Smaller places, like a private residence, may only need two or three investigators.

II. Respect private property. Not all cemeteries may be accessible after dark and not all haunted looking places welcome uninvited ghost hunters. Always obtain permission to be where you are. Getting arrested for trespassing is something you should actively avoid.

III. I don't care who was supposed to run the wire to the IR camera back to the DVR set up. If it didn't get done, don't scream about it. Act professional. People will watch you on an investigation and your conduct will speak for all of us. If you act up, throw things around, yell and bark orders like a drill instructor you make yourself look bad...and every other ghost hunter as well.

IV. Please, please, please do not call your EMF meter a ghost meter. I think this is self explanatory.

V. Leave the alcohol and drugs at home. You don't need to use anything that is going to alter your perceptions.

VI. Don't smoke on an investigation unless you coordinate smoke breaks with all other team members. And that goes for people watching the investigation. The last thing you need is phantom cigarette smoke ruining your pictures.

VII. Your cell phone is going to interfere with something on an investigation. Leave it off and at a central staging area where it is not going to interfere or interrupt your EVP sessions.

VIII. Respect the spirits you are dealing with. These were once living and breathing human begins just like you. For some unfortunate reason they are here. They may lack a physical body but nonetheless treat them with dignity and respect. Rule of thumb: treat any ghost as you would want a ghost hunter to treat the spirit of someone you have lost.

IX. Ghosts are not pets. If you have a ghost you should not want to keep it around. They are not there for your entertainment or amusement. They are not something to be bragged about over dinner. Again, we are dealing with once living people. Treat them as such and encourage them to move on.

X. Do not get hung up on debunking or other fads in ghost hunting. Do good work every time that you investigate a haunted location. Be open-minded and look for evidence and be open to experiencing the haunting as well.

XI. If you believe that personal experiences are invalid as evidence, please re-read that chapter in this book! You are missing out on a big part of paranormal research and you are perpetuating the myth that personal experiences are not reliable.

XII. Just don't settle for finding out evidence of a haunting. Try to find out the 'who' and the 'why' behind it. Maybe the ghost simply needs his or her story told. You can be their voice from the other side and tell that story.

XIII. Just because you don't find any evidence or have any experiences at a location does not mean that it is not haunted. Places with a rich history of reports may warrant more than one investigation.

XIV. Please do not tell people that ghosts are safe to have around you. They may not be. I have been on investigations where the ghosts have all but consumed the personalities of homeowners. The ghost has literally taken over in some cases. Not in a demonic possession way – more of how the negative energy of the ghost wreaks havoc upon the living. People report being scratched, shoved, and pushed by spirits. I have been pushed down several steps and had my head shoved towards a window during different investigations. I will never tell anyone that hauntings can not be harmful. Neither should you.

XV. Respect others in the field. There are a lot of ghost hunting groups out there and each has different titles and protocols. What works for you may not work for me, and what works for me does not work for someone else. There is room for all of us in the field. As long as you're professional, credible and seek the truth then we are all on the same page.

XVI. Leave the Ouija boards at home. Don't bring anything that could give your hosts the wrong impression.

XVII. Never, ever lie or exaggerate evidence. If you are not sure or there is any doubt as to the authenticity of evidence dismiss it. Skeptics love to make a mockery of the field by parading faked photos or misinterpreted evidence about. Take your time and become good at evaluating your evidence. When you are not sure about something label it as such and ask others what they think. Discussion about whether something is paranormal is better than being mocked over something that obviously isn't.

XVIII. Safety is a big concern. Do not go anyplace alone. When investigating old places watch for dangers caused by decaying buildings and potential trespassers. The living may be of concern at night outdoors at cemeteries and other locations.

XIX. Respect the privacy of the people who live at the places you investigate. Only talk about the things that you have received permission to pass on or publish.

XX. Have fun. For most of us ghost hunting is something we do because we enjoy it. Being professional doesn't mean that you can not have a good time while you are ghost hunting. Taking breaks – especially during intense investigations – is important and a bit of humor and lightheartedness can help ease the tension.

A moment of fun during a break in an investigation. I am pushing Linda about on an old baggage cart in the former lobby of the abandoned Tamarack Lodge in Greenfield Park, New York. Photo courtesy of Barbara Bleitzhofer.

Dangers that we face! Here you see how a decaying building at the Tamarack Lodge poses more danger than any of the potential spirits that may be stuck there.

Good ghost hunting is the product of professional conduct, attention to detail and plain old common sense. This is a growing field of study and protocols and techniques are bound to change and be modified over the years as we find what works and what doesn't. Take your time, be yourself and always strive to be the best ghost investigator that you can be. Ghost hunters – and their skeptics – will come and go. But one thing will remain certain. Ghosts will continue to fascinate our imaginations and fuel or passions to find them and the proof of their existence. For me, the proof is in the believing. I have seen them, felt them and been touched by them. Now it is your turn to get out there and do the same. Good luck and happy hunting!

ABOUT THE AUTHOR

Michael J. Worden has a unique background, having studied and worked in the field of exercise physiology for over ten years before pursuing a life long dream of being a police officer. He has been a police officer for ten years, a detective for the past three and a half. Michael combines his background in the sciences with his investigative knowledge and experience with great results. He has a life long interest in the paranormal and has been a paranormal investigator since 2000. He works with noted author and paranormal researcher Linda Zimmermann.

Michael is the proud father of twin sons who enjoy hiking, exploring, fossil hunting and of course, ghost hunting, with Daddy. When he is not investigating the living, or the dead, he enjoys relaxing in the sauna or hot tub as well as listening to music, writing and home brewing beer. Michael loves to travel and has made numerous trips abroad where he has fallen in love with the Scandinavian countries. He is particularly fond of Finland and some day plans on a summer retirement residence there.

ABOUT SISU BOOKS

Sisu Books is an independent publisher of books on unique and specialty topics. Sisu Books is proud to debut *Ghost Detective* as their first title in print.

The word sisu is a Finnish word meaning determination, strength of will, perseverance, mettle and sustained courage. For additional information on Sisu Books visit www.sisubooks.com or write:

Sisu Books
PO Box 421
Sparrowbush, New York 12780